ROPE RESCUE FOR FIREFIGHTING

ROPE
RESCUE
FOR
FIREFIGHTING

Ken Brennan

FIRE ENGINEERING®

PennWell
MEDIA FOR STRATEGIC MARKETS SINCE 1910

Copyright © 1998 by Fire Engineering Books & Videos,
a Division of PennWell Publishing Company.

Published by Fire Engineering Books & Videos
A division of PennWell Publishing Company
Park 80 West, Plaza 2
Saddle Brook, NJ 07663
United States of America

James J. Bacon, editor.
Book design by Max Design.
Cover art by Steve Hetzel.
Cover photo by Ron Jeffers.
On the cover: Lieutenant Cory Parker, Fort Lee (NJ) Fire Department Rappel Team.

1 2 3 4 5 6 7 8 9 10

Printed in the United States of America

Library of Congress Cataloging-in-Publication Data

Brennan, Ken, 1963-
    Rope rescue for firefighting / Ken Brennan.
        p.   cm.
    Includes bibliographical references.
    ISBN 0-912212-61-6 (softcover)
    1. Lifesaving at fires.   2. Rope.   3. Rappelling.   I. Title.
TH9414.B74   1998

628.9'2—dc21   98-2756
                                                       CIP

About the Author

Ken Brennan entered the fire service in 1978 as a cadet firefighter with the Levittown (NY) Fire Department. Since that time, he has worked in 12 different fire stations in the municipal, military, and industrial environments. He is a state fire instructor in both New Hampshire and Maine. Brennan currently serves as the operations and training officer for the City of Dover (NH) Fire and Rescue Service.

Dedication

To my wife, Kathleen, who endured while I was pecking out this book letter by letter on the computer.

To my nine-year-old son, Timothy, who thinks he wants to write a fire book himself. I wish him well.

Acknowledgments

For me, personal acknowledgments are almost impossible to set down on paper, since many people have shown me the way and have influenced my fire and rescue knowledge. I was virtually a youngster when I began my career, and I started out with a bang of sorts. I was sworn in on a Tuesday night, and 24 hours later I was on my way to fire school for training. I stood outside a six-story burn building with a rescue company, waiting for things to happen. They took off my blue helmet and put on a black leather one in its place. They said, "Hoseline or search and rescue?" and I said, "Search and rescue!" Then five of the six floors of the tower were set on fire, and the adventure began. We conducted our work from the fire escape that led to each of the windows. I was taken floor by floor through the rooms of flames, searching for the simulated victims. We met the truckies on the roof while the engine company pushed its way to the sixth floor. I received some minor burns on the back of my neck, for which my father yelled at me; however, as they say, the rest is history!

I would like to thank the following people and organizations for assisting me along the way:

Lieutenant John Barotti and Chief Kevin Sutch (deceased) of Rescue Co. 4, Levittown (NY) Fire Department—for that first fire.

Assistant Chief David A. Clement (retired), USAF Plattsburgh AFB, New York—for allowing us to stretch the rules a little bit.

Chief Thomas Brennan (retired), Waterbury (CT) Fire Department—for giving me my start in publishing while I was in Germany, and for asking, "What the hell is a carabiner, Kenny?"

I would also like to thank a few individuals who read through my drafts and shared some advice:

Battalion Chief Marty McTigue, FDNY.
Captain Dave "Mac" McLean, City of Dover (NH) Fire and Rescue.
Lieutenant Ray Lussier, Auburn (ME) Fire Department.
Firefighter Don Preble, Old Town (ME) Fire Department.

Additionally, I would like to thank the following organizations for photographic support during this project:
Chief David F. Bibber and members of the City of Dover (NH) Fire and Rescue Service.
Chief Paul Vallee and members of the City of Somersworth (NH) Fire Department.

Contents

Preface

A wide variety of situations may require fire and rescue services to perform a rope rescue. Operations on tall buildings, towers, or bridges; below grade; in confined spaces, swift water, rugged terrain, and on ice all require the systematic use of lifesaving rope techniques. The ever-increasing demand for such services increases the possibility that firefighters may be injured while performing them. The best way to manage these increased risks is to train firefighters properly in rope rescue operations and to follow good safety practices. This book is a compilation of many thoughts, ideas, and practices put forth by various authorities. In it, I will try to give the basics to the reader, covering both the tangible components of rope systems and the intangible concepts relating to them. Within every area of human endeavor there are differences of opinion as to how to get a given job done.

In the discipline of rope rescue, particularly in the United States, there is a divergence of philosophies. Many times, regional influences play a large role in how specific techniques are done. The issue of what piece of hardware or software to use can create interesting discussions. Having several techniques and different types of equipment for a given situation is critical in the event something doesn't go as planned.

Still, every rescue team sets out with the same purpose in mind—to save a life. It's how they accomplish that feat that perhaps best distinguishes the individual approaches. To be safe, efficient, and successful are the governing parameters that apply to us all.

1

Chapter One

Assessing a Community's Needs

To assess means to size up or evaluate; to weigh whatever variables are present so as to most efficiently handle an incident. This includes the assets and liabilities in whatever jurisdiction you happen to protect. An assessment can be broken down into four subcategories: topography, construction of the occupancy, demographics, and in-house capabilities.

TOPOGRAPHY

When you think of the word topography, what comes to mind? Many envision rugged terrain or wide-open spaces. Basically, topography refers to any existing characteristics of an area; the physical features of the landscape, natural or man-made. Every jurisdiction has distinguishing physical features that present both positive and negative aspects when a rope rescue is necessary.

From a community development view, you can think of a jurisdiction as a single entity comprised of linked population centers. Sections may be wilderness, rural, suburban, and urban. Many jurisdictions are comprised of two or more regions of different types. The one I work in, for example, is comprised of urban, suburban, and rural areas.

Suppose your community is urban, developed, metropolitan. Look past the steel and concrete, and focus instead on its raw physical features. The need for rope rescue techniques in such environs may seem remote, but change your mindset to reassess the threat. Going over past run reports may help toward establishing a need for specialized training. Consider, for example, a scenario in which a car goes over a guardrail on the local interstate. On your arrival, the car is down a 100-foot embankment. There is a small river below with limited marshland

Rivers and minigorges are common features in many industrial communities.

access, and you can't use a tower ladder. Although the general topography is flat, the downtown area is split in half by a river that leads to a minigorge well below the street level—a common feature of many New England communities. Such a scenario, even in the middle of a city, can be vexatious to many departments relying on traditional

A typical sand pit.

options. Your first instinct is to call for an established rescue team; however, that takes time, and you desperately want to do something. (It's in a firefighter's blood to act no matter how dangerous the situation.) It has been said many times that rescue is matter of transportation—reaching a victim in a bad situation and getting him out of there. Topographic features, even man-made ones, can greatly affect the character of any proposed rescue mission.

When you assess the topography in any given area, some of the features you should evaluate are:

1. Waterways, including rivers, lakes, ponds, aqueducts, sumps, cisterns, and storm systems;
2. Terrain eccentricities such as hills, slopes, plateaus, and the like;
3. Subgrade features such as quarries, sand pits, and dry river beds; and
4. Vegetation, including the relative density of the plant life, plus the quality of the soil or water underneath it.

Each of the above characteristics poses its own transportation problems. For those who believe in Murphy's Law, you know that there are going to be additional facets to any operation. Here are some factors to make your day:

1. Your victim is well off the road.
2. You arrived with a small contingent of personnel.
3. It's raining, and the temperature is dropping.
4. It's near shift change, and it's getting dark.
5. Your capabilities are limited, and you haven't reached the victim yet.

Lifesaving rope techniques can greatly enhance the survivability of the victim. Proper use of rope can get a rescuer to the victim faster and make rapid medical intervention possible. It can mean safe and rapid extrication from a hostile environment. Traditionally, regional search and rescue teams have handled such incidents in areas that are predominantly rural or wilderness. It's these organizations that have influenced the fire service so dramatically by developing the equipment and techniques used in cities and towns today.

CONSTRUCTION AND OCCUPANCY

Construction

For purposes of rope rescue and firefighting, construction features can generally be broken down into five types:

Type I:	Fire resistive		**Type IV:**	Heavy timber
Type II:	Noncombustible		**Type V:**	Wood frame
Type III:	Ordinary			

A water tower that has been used by graffiti artists.

A vertically lifted drawbridge.

What do these five types mean in the real world? In my region of the country, the multistory and high-rise structures are fire resistive and noncombustible. The classic downtown brick-and-joist structures are

A transmission tower.

of ordinary construction. The mill and old shoe shop structures are of heavy timber, and the homes are predominantly wood frame.

None of these structural types mandate rope rescue procedures; however, any building might become the site of a rope rescue during periods of construction, renovation, or demolition; from the effects of storm damage, earthquake damage, fire, and many other calamities. Although construction features are important, it's the combination of construction, occupancy, and circumstance that creates the essential hazard. The potential hazards increase in and around special structures, which can be part of, on, or separate from buildings of more conventional design. Some examples are TV and radio towers, bridges, and confined spaces. This last category is nearly endless and might be any space that a person can get into but that wasn't designed for human habitation. Such a broad definition encompasses everything from the below-deck areas of vessels to agricultural silos to HVAC duct systems to pipelines to underground industrial facilities.

Occupancy

The word *occupancy* generally denotes what kind of business or activity is going on in a given building. When sizing up a structure, differentiating between a school, a supermarket, an industrial plant, and a storage facility is critical. It's generally the nature of the business and the

Even an ordinary steel bridge may be the scene of a rope rescue.

physical layout of the building that contribute to the specific hazard.

When you think of a six-story heavy-timber mill that is occupied by white-collar office workers, you wouldn't ordinarily envision rope rescue services being required in the event of a crisis. Suppose, however, that the rear of the building is abutted by a river. The building experiences a gas explosion in a service area that houses a sprinkler riser, rendering it ineffective. A fast-moving fire traps several workers against the rear fifth-floor windows. In this instance, aerial apparatus can't be positioned for the rescue. What tools does the fireground commander have readily available to deal with this situation? After quickly running through his options, he orders a qualified team to the roof to effect a rope rescue.

An ordinary steel bridge spanning a river can present similar problems. Thrillseekers are known for climbing in and around bridges just because they're there. Bridges also attract those with suicidal tendencies. Dealing with a suicidal person on level ground is difficult enough, never mind being perched high above water. Whatever the specific situation, the safety of the personnel operating near or directly with the victim is paramount. The first hurdle to get over is in establishing eye-to-eye contact to help that person out of his predicament; the second hurdle is to remove that person without getting yourself killed. Rope techniques are essential in our world, and if you undertake a rope rescue without the proper training and prerequisites, you are inviting catastrophe to the operation. The human element only increases whatever liability is presented by gravity, fire, water, and other forces of nature.

Certainly industrial environments present great potential for rope rescue. Complex machinery and exotic processes pose as high a degree of danger as the worst special structures. Tunnels, vessels, vaults, cable trays, vats, scaffolding, and a myriad of other industrial features require maintenance, inspection, and testing. Municipal fire depart-

A myriad industrial features require maintenance, inspection, and testing.

ments are commonly listed as the primary rescue service for many industrial facilities in their confined space rescue plans, often without the department's even being aware. Disaster can easily attend such scenarios. According to 29 CFR 1910.146, *Permit Required Confined Spaces Regulation,* rescue services must be incorporated into the written permit system, and they must be ready to respond and operate in a relatively short period of time. Rope rescue techniques and equipment *do* work well in confined space operations. These facilities should be prime reasons for inaugurating a rope rescue team in your district or even for upgrading your present capabilities.

DEMOGRAPHICS

The word *demographics* refers to the statistical study of human populations. Such studies look at numerous data, including size and density of the populace, their distribution, purchasing habits, and many other traits. This information can be useful toward determining departmental needs.

Demographics, as I view it, is a broad gathering of information on human activities; a breakdown of a population's age, income, leisure activites, education, and the like. Such information is available from many databases. Some common sources include:

1. The building department can tell you the number of permits for construction that are outstanding and under request.
2. The tax assessor can tell you the number and types of structures in the area, which will allow you to assess population clusters and thereby to target hazards.
3. The planning department looks at future development at the macro level of the community. Generally, more buildings equals more people.
4. Economic development also looks at the big picture and can tell you who has received and who is applying for block development grants.
5. The chamber of commerce works as a marketing organization for the community and seeks to assist established companies as well as emerging ones to locate within a given area.
6. The board of education can give you current and projected enrollments. Any increase can affect the community tremendously.

7. Utility companies are helpful in assessing communities. Their bottom line is to sell a service, and if it's not there, they have to construct the appropriate delivery system of power lines or pipes. Demand for new or expanding service is a strong indicator of growth.
8. Federal and state OSHA agencies collect data on worker injuries.

These are just a few sources to get a demographic profile of your community and that are just part of the assessment matrix. As a general principle, as the population increases, and as facilities are built and occupied, the risks to civilians and firefighters naturally increase as a consequence.

A progressive fire prevention bureau would have much of this data if it has performed a life safety strategy survey. Collect data on the types and numbers of occupancies—i.e., the numbers of single- and multifamily residences, commercial and industrial properties, special structures, and other high-hazard facilities.

The matrix can be as simple or complex as you care to make it. This gives you an idea where you came from, where you are, and where you're going. The following hypothetical analyses are suggestive of different levels of need for rope rescue training:

Analysis One: If your jurisdiction is predominantly elderly and contains virtually nothing but single-family residences, it can be viewed as a low-risk area. Demographically, such population groups have a low need for rope rescue services.

Analysis Two: Your jurisdiction has a large population of transient younger people aged 17 to 24. A further look at EMS reports, coupled with an increase of activity in the adjacent wilderness area, denotes a trend. Your final analysis shows a large jump in enrollment at the local university, as well as an increase in drug- and alcohol-related incidents. Young people happen to get injured out in the woods, and simple EMS runs become more complicated when rope rescue is needed for extrication.

Analysis Three: For years, your bedroom community was quiet, and its growth was limited to residential structures. Then, as in many communities, businesses and industry found their way to your town and began to build. Regardless of the reason they found your district desirable, the newcomers are now your problem. The impact of a sudden population growth on services can be tremendous, sometimes overwhelming. Many jurisdictions can't absorb the financial strain and so deal with it in two ways. First, they can tack impact fees onto construction permits; second, they can require off-site exactions of new corporations. Examples of such exactions include constructing roads and traffic lights, building sewage treatment plants, and purchasing new fire apparatus for the community. Such indicators are useful in evaluating your department's need for rope rescue capability.

Analysis Four: A large tract of wilderness land is turned over from the military to the local state government. This land transfer is part of the downsizing of the U.S. military and is in the interest of the public's recreational use. This area contains rivers, lakes, ponds, and several mountain peaks. These peaks quickly become popular with sport climbers because they were formerly restricted to military personnel only. The local community is now responsible for rescue services for anyone who gets lost, trapped, or injured in those areas. Combining the resources of the fire department with a wilderness search and rescue team will be of benefit to all.

Analysis Five: A newly discovered series of subterranean passages gets the attention of cave explorers. Everyone from the novice to the expert wants to experience this beautiful new find. Hidden among its wonders are many dangers that may prove deadly to the uncautious or ill-prepared. Dealing with darkness, water, cold temperatures, potentially hazardous atmospheres, uncharted routes, and a wide variety of passage sizes can take its toll on explorers. Even a minor problem can require a major rescue operation. This would entail getting rescuers to the victim, providing any medical assistance and packaging for transport, then getting the entire contingent back to the surface again and the victim to the hospital. Very few agencies are either equipped or have the training necessary to accomplish this type of operation. The local community must contact the nearest cave rescue organization and review the options long before the need for them arises. Ultimately, once a caving incident occurs, the nearest trained cave rescue team will respond to handle the problem. This may be time-consuming and tedious; however, it's far better to wait for the experts than to endanger personnel who would be wholly out of their realm.

Analysis Six: A series of worker injuries at a paper mill has prompted an OSHA investigation of the overall plant safety program. Many of the injuries have centered around fall protection practices or the lack thereof. The fundamental problem is that the workers aren't using any safety equipment when operating on or around the various building components that pose a fall risk. A consultant is hired and the proper training is given; equipment is purchased and procedures are written. The last element to be added is that of rope rescue services. Safeguarding an employee is one consideration; providing contingency options for an employee who has taken a fall is another. The employer must not only protect the employee from the fall, he also has the responsibility of getting that worker removed to a safe location. Here again, the company is going to look to the fire department for this service. Hopefully the company will actually inform the department to initiate a planning process, not just pencil-whip it for insurance purposes. The local community may or may not be trained or equipped to deal with this sort of situation. Combining the efforts and resources of plant and department personnel alike is the best and likeliest starting point.

You see how topography, construction, occupancy, and demographics can tie together. You need to look at the major tasks that the community's employees perform. Set up a risk analysis for the major employers in your district, and look at the hazards associated with the various occupations. Examine the jobs or processes involved, and see how rope rescue services might be required. Statistics are available from OSHA and your state equivalent, such as the Bureau of Labor Standards, if your state operates this way. Such organizations keep data on injury and death rates of specific high-hazard industries, including logging, construction, manufacturing, transportation, the marine trades, mining, and others. Your state labor department should have data on workers' injuries and the dispositions of their cases.

IN-HOUSE CAPABILITIES

These are the capabilites that the fire organization already has in place to offer its customers—yes, customers. Whether you want to admit it or not, the fire and rescue service is a business. We have budgets and answer to taxpayers, we provide unique services, and everything we do today comes down to issues of dollars and cents. On the macro level, every fire department provides rope rescue services in one way or another. Unfortunately, for many, that doesn't mean anything more than a rope tied around someone's waist to pull him out of a hole or lower him off a roof.

This section on assessment starts with listing your present skills along with equipment supplies. Evaluate how many personnel are trained and ready to provide rope rescue in a reasonable period of time. As the types and levels of service increase, so too do the required skills, equipment, and training.

Generally speaking, there are several ways to break down the services provided. Some groups list them according to the plane of operation within which they occur:

1. Be able to descend and retrieve a victim.
2. Be able to ascend and retrieve a victim.
3. Be able to operate horizontally and retrieve a victim.

In reality, providing rope rescue services is more than tying a bowline around your waist with hemp. For our purposes, the acceptable minimum is that personnel will be trained to one or more of the above-mentioned categories. Also, all rope, harnesses, and hardware must be compliant with NFPA 1983, *Fire Service Life Safety Rope and System Components*, 1995 edition.

Once you've assessed your department, take a look at other agencies in your jurisdiction that may provide rope rescue services. In certain areas of the country, law enforcement agencies such as county sheriffs departments have this capability. Often this is because their areas are largely rural and they have the authority to operate in both incorporated and unincorporated environments. You may want to regionalize with them to pool your resources and expertise.

Another intercommunity agency that you can assess is the local civil defense unit. In some communities, local civil defense units provide rescue services or EMS to the community.

One other local agency that comes to mind is a wilderness search and rescue unit. These units can be privately run or of a quasigovernment structure. The government side usually gives them the authority to operate plus partial immunity from liability. Agencies such as these can assist you in many ways, and it's important to establish a rapport with them. Regionalizing can be a catalyst to forming a truly viable team.

There are those who categorize rope rescue services even further, splitting them into fireground and nonfireground (technical) services. The bottom line is that qualified personnel must arrive with the first-in fire companies to effect a fireground rescue. The nonfireground rope rescue capabilities are equally important; these, however, can be performed by personnel outside of the department. The reason for this further definitional breakdown is to establish that nonfire personnel can't operate on the fireground. You need to assess how this will affect your operations. Being able to understand your capabilities and train enough firefighters to do the job is where topography, demographics, construction, occupancy, and other factors tie together to form the big picture.

When the Phoenix Fire Department got into rope rescue, it set itself up for an urban environment. As time went on, however, it realized that a majority of its incidents were occurring in two popular parks in the city.[1] Modern techniques are quite adaptable to most settings most of the time. Your organization must be equally flexible and dynamic.

Assessing the community is a constant process that must also be continually reevaluated. I hope that this chapter will give you some insight into what's needed to get the ball rolling. When you make your presentation to the chief or ask for a bigger budget from the city council, you will need to have the facts and figures to shore up your position. In many medium-sized cities, department heads and program managers have to explain their requests in front of televised council meetings. In smaller towns or in joint funding districts, requests for new or expanded funding are presented to boards of selectmen or committees. Those who appear unprepared often fall victim to scathing attacks that can get personal at times. Be prepared to be your own first rope rescue statistic!

STUDY QUESTIONS

1. An assessment of a community's needs can be broken down into what four variables?

2. Name the five types of building construction.

3. To what does the term *occupancy* refer?

4. What OSHA regulation covers rescue response in confined spaces?

5. The statistical study of human populations is known as _____.

Chapter Two

Organization

In Chapter One, we looked at factors that influence the need for life safety rope techniques. In this chapter, we'll look at the nuts and bolts of organizational structure, training, and the incident command system as they pertain to rope rescue services.

The term *operations* refers to the hands-on activities of an incident. When incorporated into an incident command system, the operations section is responsible for the tactical activities needed to meet strategic goals. As a quick review, strategy pertains to the overall goal that the incident commander wants to accomplish, whether it's to put out a fire, save a trapped victim, or protect an exposure. Tactics are those small enabling steps that, when carried out properly, help to achieve the strategic goal. Tactics include placing ladders for rescue, stretching hoselines, and setting up anchor systems. Each incident that you handle can be termed an operation or, as it is traditionally known, a job. This term in recent years has expanded beyond the fire incident and is now used to describe other emergency incidents as well. The term *operations* means different things to different people; however, for our purposes, it's the all-encompassing theme for running the team. For an operation to be successful, it must be part of the larger concept called the *organization*.

We all need to belong to an organization. Without one, bedlam and chaos would prevail. Safety, effectiveness, and efficiency are the by-products of good organization. In reality, we all know that having a group of people together sharing a common goal doesn't guarantee that things will run smoothly. For the sake of argument, we'll establish the organization as a positive step. Organizations exist before, during, and after any given incident. The local organization, be it municipal, regional, volunteer, military, or private, is structured and empowered in advance of need. This larger jurisdictional unit is called the host or parent organization. It's this parent that gives us the legal authority to perform our duties. In most cases, it's the local fire department, search and rescue group, military unit, or industrial brigade that funds, trains, supports, and uses the team's services. Every team must establish its own line of authority to function.

When an incident occurs, the first-arriving officer sizes up the situation and determines his needs. These needs guide him toward mobilizing a technical or fireground organization to deal with the problem.

In the past, many organizations didn't want to perform any rescue service activities. Today, many are doing just that for a variety of reasons, including political necessity and to maintain their piece of the budgetary pie. No matter the incentive, the following depicts a variety of parent organizations that provide these services. These are organizational structures, not ICS flowcharts. The purpose of showing the organizational structures is to highlight how similar-size departments provide these activities.

In smaller departments, specialty services such as rope rescue are the result of collective effort. Typically, a majority of the on-duty shift get involved. Equipment is typically on a rescue, utility, squad, or aerial apparatus.

A typical combination department (part paid, part volunteer) staffs a career duty company with one officer and three firefighters. They staff one engine and cross-staff the ladder and rescue apparatus. In this case, EMS is handled by a third agency. Additional personnel are

Small Combination Department

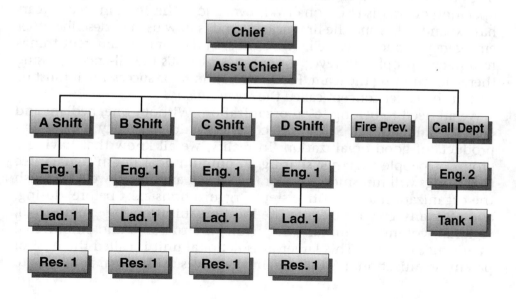

Small Career Department

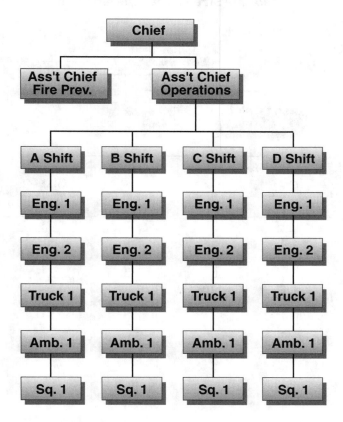

recalled by pager. These members may be both career and call personnel. During the daytime, the chief officers are available.

When the duty shift gets committed to a rescue, it strikes a box alarm as a call for station coverage and/or to call additional personnel directly to the scene.

Author's note: Due to shrinking budgets and increasing demands for service, the traditional role of staff personnel is changing dramatically. This is especially commonplace in smaller and medium-size companies. These personnel not only advise the line personnel, they also work and respond with them. Such creativity is essential to surviving in today's governmental environment.

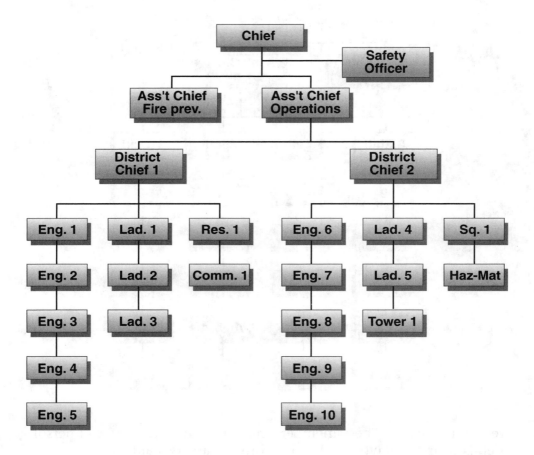

Medium-Size Career Department

In medium-size departments, specialty services such as rope rescue are provided through a multicompany effort. Several truck companies, espcially those that cross-staff heavy rescue apparatus, are typical providers. Departments in this category can cross over and become full-time staffing of a rescue company. When this occurs, they become the primary providers of rope services.

Medium-Size Volunteer Department

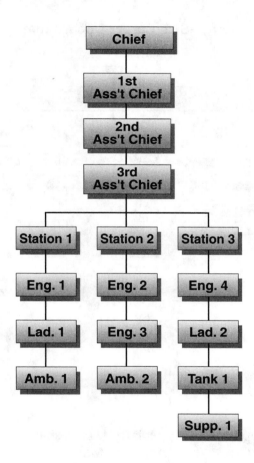

The accompanying diagram shows a medium-size volunteer department, typical of the East Coast, serving a large suburban area with commercial and industrial development. The ladder companies provide basic extrication services, and the rescue company provides more advanced capabilities.

Large-Size Career Department

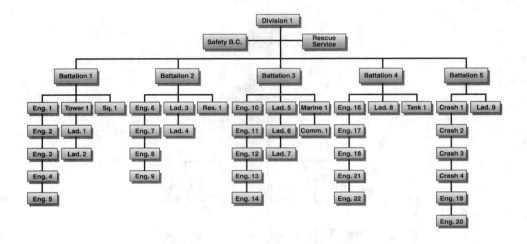

In large-size departments, specialty services such as rope rescue are provided by heavy-rescue companies. Many of the larger metro areas in the United States staff between two and five rescue companies, several tactical support companies, and a rescue service supervisor/coordinator, all of whom respond to multicompany operations.

Regionalized Team

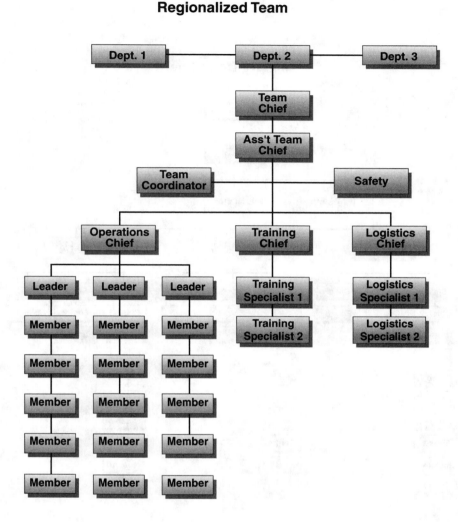

Regionalized teams are formed by the joint effort of two or more departments. Typically they provide the necessary staff, funds, training, and facilities, the cost of which would be prohibitive if done by any of the entities alone. This team usually has established communication and standby procedures so as to facilitate a timely response to incidents.

Military Department*

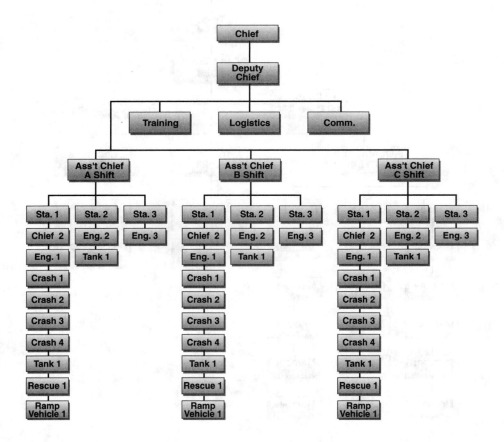

*Typical active-duty departments work a 72-hour schedule in two-shifts. National Guard units work a three-shift schedule.

Military units that provide fire protection normally work under the small-department setup. Three- and four-member rescue companies facilitate the rescue while additional suppression personnel provide direct support.

Large Industrial Brigades

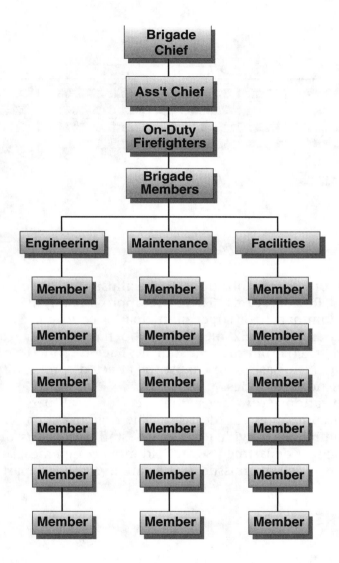

Large industrial departments are made up of two components: They have on-duty firefighters and are backed up by on-call brigade members. The primary rescue services are provided by the on-duty firefighters and supported by the brigade. In some instances, select brigade personnel are also trained to the level of the on-duty firefighters.

FEMA Urban Search and Rescue Task Force

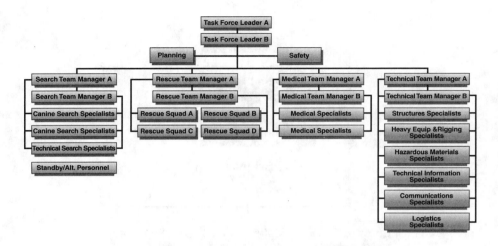

The last of the organizations to be illustrated are FEMA's Urban Search and Rescue Teams. There are approximately 25 teams, comprised predominantly of larger municipal or county fire departments. Staff levels are set at 62 members, who have certain response and equipment criteria that they must meet. They respond to major natural or man-made disasters and assist the affected community with specialized rescue capabilities.

As organizations grow larger, the need for recall personnel and the number of companies can decrease. The organizational charts reflect this and omit the individual company staff particulars due to sheer numbers. Larger departments can field more companies, since personnel aren't as much of an issue as they are in smaller departments.

THE ROPE TEAM

The actual rope team that handles an incident can go by different designations, such as high-rise team, hazardous response team, special operations team, emergency services unit, rescue services, high-angle team, rope rescue team, and heavy tactical rescue team, to name a few. Teams are structured similarly to parent organizations with regard to the various stages of an incident. The difference as to how rope rescue is carried out at this level depends on numerous factors, such as community assessment and inventory of liabilities, the parent organization's directions, the complexity of the incident, funding, training levels, equipment, other teams in the area, and the like.

When you think about the components of any evolution, one overriding principle to keep in mind is safety. Attention to safety within the parameters of an operation implies a fine balance between risk and benefit. Usually the resultant benefit of an operation is to save a life, whereas risk is typically the danger posed to the firefighter. In practical terms, safety and safe practices affect our lives every moment of every day. What is safe to one may not be safe to another, since each situation poses different hazards, the nature of which will never be quite the same from one instance to another. In a utopian world, I suppose that risk and benefit would be of equal proportions; however, we don't live in a perfect world, and the scales are generally tipped heavily toward the risks. This is another reason each officer needs to have a good grasp of the territory in which he operates. Eventually you will have to make a split-second fireground decision that may mean accepting a risk only to find that the victim to be rescued has already expired. In such a case, prudence might have been the better course.

The primary impetus for stating this theme is the current debate over something that is probably indefinable: what is safe and what isn't. Safety itself is paramount, and the lack of it can cost an untold amount of grief and human suffering. The very nature of our profession is hazardous, since we court many risks and must continually, individually manage them. Managing these risks can be reduced to the following equation:

Proper training + proper equipment + thorough assessment + correct use + safety practices + close supervision + a little luck = a manageable risk.

Looking at each component separately, it's easy to see that each incident presents these criteria in varying proportions. The individual team member must have an appreciation of the components of this operating matrix. Many times an incident unfolds rather quickly, and the matrix comes together instantly. Other times, a more tedious progression is involved. However the growth of an incident, having met these criteria beforehand as much as possible will lessen the risks. Purchase the right equipment, train all personnel to the appropriate level, enforce safe procedures, initiate standard operating guidelines, and establish a command structure. Only then can you routinely minimize the risks.

As mentioned earlier, rescue is a transportation problem, being first a matter of getting to the victim, then removing that victim to a safe location.[2] Safety and transportation concepts tie in directly to how big a team and its organization ought to be. In the search and rescue business, the acronym LAST stands for locate, access, stabilize, and transport. Many times, the locating has been done by a passerby or other civilian. Each situation is different, however. Organizationally speaking, it may take several hundred members to locate a victim in a wilderness area. You can break down the operation in terms of the skills required to effect the rescue. Once you have located the victim and decided to perform a rope rescue, the next step is to deploy those personnel who are trained in the appropriate skills. The nature of the emergency will dictate how many members will directly participate in the mission.

Generally, you cannot have a team with fewer than two people, although, technically, a single person could effect a successful rescue. This is where each team will disagree and set up operations according to its own standards. Some teams take the buddy system approach, which means that two people are assigned to each critical area of the total system.

To better understand this, one approach dictates breaking down the technical rescue into three parts: the anchoring system, the access system, and the transportation system. Imagine the scenario of a cliffside extrication using a double-rope configuration. To ensure an efficient life safety system, all three subsystems must be in place. A task breakdown could be as follows: One person (plus a buddy backup) sets up a hauling system, one person (plus buddy) sets up to rappel to the victim, and the last (plus buddy) sets up the stokes for a horizontal haul. Each buddy member watches over his counterpart to ensure safety and redundancy. Another approach to using those same six is to have both members in each subsystem work together to build their particular component. One of the major factors that will guide this operation is the experience of those involved. It is also important to recognize who is in command and who is responsible for safety. Including these two responsibilities, we now have an eight-person team of specialized members, not including any other support personnel who may be involved.

Let's break down this scenario in terms of the acronym LAST. Locating the victim was apparently done by someone other than a team member—i.e., a civilian. The access is cliffside and relatively easy with respect to moving equipment to the site. Stabilization begins as soon as the first rescuer descends to the bottom and establishes the patient-caregiver relationship. Transport is by way of a stokes basket being hauled to the top of the cliff. Obviously this is no small undertaking; however, that eight-person team could have been handled by four trained members plus several support personnel. Would it have been as safe or as smooth an operation? Each organization must take a hard look at what services it is going to provide and at what level.

Imagine now an incident in the woods of a local park. A victim has slid off the trail and has sustained life-threatening injuires. There are no possible helicopter landing zones for one mile, and the nearest passable road is even farther. Breaking down LAST in this instance: The victim was located and the alarm was initiated by fellow hikers. At least one hiker is guiding you to the site. In assessing the situation, the team will have to haul everything they need on their backs through the woods from the road, two miles away. Hopefully, established trails will make access simpler, thereby eliminating the need for bushwhacking. Stabilization will begin once the first rescuer gets to the patient. This may be delayed by whatever packaging is necessary for the arduous transport. Once rescuers have hauled the victim to the top, they can transport him with a stokes carry. How large a team would this sort of operation require? Technically speaking, anyone directly involved with the operation is a team member. Each contributes to the outcome. For our purposes, a team member will be defined as someone who is trained in rope rescue skills. A support member is someone who provides other skills, up to and including brute strength. Returning to the question of the size of the team: You need to expedite the victim's transport, and that's being done by sheer brute strength. It will probably require at least eight members just to get the victim out to the road. Additional members have to stay at the site to clean up and remove the equipment and perhaps to assist with any investigation of the matter. Barring any significant weather or darkness problems, the team could be of from six to eight members, with a dozen other personnel acting as support.

TRAINING

The training process has the greatest impact on emergency operations, more so than any other facet in the organization. It has taken many years for this to be fully appreciated by both upper management and the civilian leadership. We can no longer afford to have a probie be shown the ropes by one of the old salts while an operation is in progress. When discussing experience, it is often asked whether a person has twenty years on the job or one year times twenty. The insinuation, of course, is whether a person has progressed in his career development or whether he has made the same mistakes time after time. You can perform a given task on dozens of occasions; however, if critical steps are flawed or missing, you aren't efficient and, worse, you aren't safe. Proper training is the foundation on which competent experience is built.

Training is a process whereby an instructor imparts specific information to students, which the students then apply to their own work. Training takes us from where we are to where we want to be. This generally means from a lower to a higher level, closing the gap between everyone's experience. For example, suppose you tell someone with little training to construct an anchor system. He takes a long time and still doesn't quite get it right. You then instruct that person on anchor selection, fabric selection, the effects of stress, and knots, and soon the results are better. You have filled in those gaps of knowledge and have allowed that person to experience the process. In training, as in other education, you are essentially looking for a change in behavior. This behavioral change denotes that learning has occurred, and it is evaluable through a testing process.

Training has become so prominent over the past decade or so for many reasons. Liability issues, increased demands for specialized services, statistical analyses of incidents, and financial considerations are only a few. There has been an increasing realization that good training fills in the knowledge gaps that internal and external influences create. In these times of doing more with less, training allows us to perform our jobs more safely and with greater efficiency.

GETTING STARTED

In almost every host department, there is someone who has had some rope training or at least an interest in receiving it. Many get started through sport climbing or the military. The host organization has to have a training division of its own, be it a dedicated facility or a one-man show. Decisions have to be made as to the types and level of service to be provided, who is to be trained, who will provide the initial instruction, and how you are going to maintain those skills that the personnel acquire. Many departments appoint a team commander or leader who is responsible for running the team. This can either be someone from the training division or someone from suppression operations. In larger departments, these leadership functions are run by individuals involved with rescue services. This individual creates an organization to manage the team. Specific management areas cover training, operations, logistics, and coordination, to name a few. This section will focus on training.

We'll call the person responsible for training the training program manager. This individual, with input from many sources, initially has to figure out how to train the new member and to do this cost effectively and with the least amount of disruption. Such a task can be a nightmare. The question of how large or small a team should be can become a major issue in terms of who is going to train your personnel and overall cost. Generally, the more personnel to be trained, the greater the cost. Do you therefore contract for in-house instruction, send your entire team away, or send away select personnel to perform train-the-trainer courses when they return? Other associated issues that come up include overtime for traveling personnel, expenses, lodging, meals, learning materials, and the like.

Looking for providers of rope instruction is like shopping for a car, since there are numerous options available. The most logical place to seek instruction is from state fire training organizations. Many probie firefighters get their initial training either from state-run programs or courses that are based on the state curriculum. A vast majority of fire service curricula follow NFPA 1001, *Standard for Fire Fighter Professional Qualifications*, which was recently revised down to two levels from the original three. This standard lists the minimum requirements and contains a small amount of rope rescue criteria, representing therefore a starting point for any team that is being formed. State and county organizations typically provide rope rescue instruction to both municipal services and private industry. The next level of training is through larger departments or established teams in your area. Other avenues are available through private companies and individuals. Quality instruction can come from any of these sources, but you need to do your homework to find the best. I can't tell you how many times I've asked a group where they went for their training and what they learned. The natural follow-up question is whether they think what they learned will enhance their operations. The typical response is that they don't know. Something was learned; however, it might not fit their needs. If you send your personnel away, they may get a canned show that is not very cost-effective. Obviously, certain basics must be grasped and mastered. The problem is in the approach that each instructor takes and in their variant philosophies.

Let's look back at our beginnings, which were in mountaineering, spelunking, and wilderness search and rescue. The practitioners of these arts pioneered the techniques and equipment that we use today. Each facet has improved dramatically, and these types of groups do provide some of the best training available. The problem that I and others have run into is that they are not firefighters; hence, they don't quite understand the needs of our world. They don't fully understand short staffing or having to deal with the variety of day-to-day challenges that the modern fire service faces. I'm not painting generalizations with a broad brush—just polishing the old maxim of buyer beware.

One thing to keep in mind is that there is no national standard or certification; therefore, curricula are based on individual approaches. What this means is that the buyer of initial training should make sure that whatever foundation is built can be expanded on. Programs can be listed in hours, by objective, or both. Many times you'll see course listings as basic, intermediate, or advanced. A 16-hour basic rescue course may encompass knots, rope construction, anchor points, hardware, harnesses, and short rappels. That's just enough to build interest without overwhelming the students. It's also enough of a dose to help weed out less-than-satisfactory candidates. On the other hand, there are also basic rope rescue courses that include all of the above except rescue knots. The knots are listed as a prerequisite and have to be mastered prior to starting the program. This puts some ownership in the individuals' training and allows for more quality time devoted to other areas. Although mastering knot tying is critical, it is also time-consuming to teach and can disrupt the flow for other students. Understand that there are many ways to provide instruction, then research all the pros and cons in whatever course your organization decides to invest in.

Here are just some of the many ways that courses may be set up, either by governmental or private training organizations:

BASIC
Safety.
Standards and regulations.
Rope construction and selection.
Rescue knots.
Hardware.
Harnesses.

INTERMEDIATE
Anchors.
Anchor systems.
Belaying.
Rappelling.
Pick-offs (window rescues).

ADVANCED
Mechanical advantage systems.
Stokes operations.
Highlines.

Some approaches divide the necessary skills between two levels, basic and advanced. Generally they teach the same topics, albeit to somewhat different standards:

BASIC
Safety.
Rope construction.
Rescue knots.
Anchors.
Anchor systems.
Mechanical advantage.
Belaying.
Rappeling.
Hauling systems.
Lowering systems.

ADVANCED
All of the basic areas listed above, in greater detail, plus some additional topics.

Pick-offs.
Knot bypasses.
Running protection.
Highlines.

Courses are available in a wide variety of scopes and depths. Some approaches eliminate some of the advanced skills altogether and concentrate on more basic ones. As they say, time is money, and both are in short supply. Consequently, it is important to analyze exactly what you're getting and how it can be expanded on later. Courses are also available according to the particular organization's preference, such as Technician I, II, III; Vertical I, II, III; and Rescue I, II, III. The NFPA is currently working on a new technical rescue standard, the preliminary methodology of which is going to resemble the haz-mat standard in terms of skill levels: awareness, operations, and technician. Bear in mind that it is important to learn more than one way to perform each skill. You can do this by using ideas from various sources. Contrary to some instructor groups and manufacturers, nobody has all the answers as to the right techniques and equipment. Just as in firefighting, there are many ways to get the job done, and a certain amount of flexibility is desirable. Many of the hardened ideologies follow a basic foundation, and integrating them in the real world is possible as long as you fully understand their operant principles. This, plus knowing how much time initial instruction should take, can generate some heated discussion among the experts. Like most of the skills that firefighters must master, there simply isn't enough time to cover all of the tasks to the degree of competence that most would like. Just to learn the skills of the basic, intermediate, and advanced lists can range anywhere from 80 to 120 hours of instruction time. Much of the rest will be taught in the real world.

Unbeknownst to many outside of the training circles, there is a war being waged on these very issues. Inside the fire service, and to a larger degree in private industry, corners are being cut to provide the necessary objectives in the least amount of time. This means seeing the skill once and moving on. Obviously, if an ill-trained student later experiences a real-world problem or failure, he may not have the knowledge or experience necessary to get himself out of trouble. There are two ways to improve the odds in such a circumstance. The first is to practice frequently back at your station or facility after the course is over. The second is to train supervisors to a greater degree of competence so that they will be better able to deal with operating difficulties.

Recruiting team members is generally easy and can at times be overwhelming when the response is particularly good. Each department needs to have its own criteria to provide for a balanced effort. Essentially you are looking for members who are physically fit, mechanically minded, and adaptable. They need to be told what is expected of them and the dangers that they will encounter. By the same token, the recruits themselves need to discuss their expectations with the rest of the team and its leadership.

In selecting a training site, establish a primary drill facility for the majority of your work. This can be an academy facility or, as is the case with most departments, a designated training area. Typically this is a facility with sufficient room for on-duty companies and that has been modified for rope rescue training. This can be as simple as using the hose tower or a ground laddering mockup. It's not so much the height you're worried about, but rather, getting the techniques right. The second selection should be of structures in your jurisdiction that are representive of those you may have to operate on. Size them up for their suitability and safety. Seek permission from the owners prior to practicing on them, of course. Real-life sites are great; however, they have problems that you must factor in prior to training on them. Many have limited access and parking during business hours, and such minor logistical faults can often cause the most headaches. Also, there is a sideshow element involved—a problem for on-duty companies that you can alleviate if you make arrangements to stage apparatus close to the drill site prior to conducting practice. You have to be careful not to create a hazard by putting on a good show that will attract bystanders and snarl traffic. Operating at a real site almost always presents a higher risk than practicing at your fixed facility, and you must make provisions for this. Sometimes the best sites are located in remote areas of a given jurisdiction, but companies may have difficulty covering their first-alarm assignments if called. If standby companies aren't available, plan your training session during off-peak hours. Most areas have peak run times, and you can factor this into your overall plan.

Climate will also affect your training activities. While the greater portion of your hands-on training will be conducted outside, sometimes extremes of weather will force you to work indoors. You can accomplish a good deal in a classroom setting, especially if it's recurrency training. There are, however, operations that demand more space. This is where you have to seek out facilities within your district that have high ceilings and bombproof anchor points. Such structures might include aircraft hangars, public works garages, large agricultural storage areas, gymnasiums, and sport arenas. This is not to say that we run for cover at the first sight of a snowflake or heat wave. On the contrary, it is important to train in adverse environments so as to ensure that your personnel can operate in less than ideal conditions.

By analogy, rope rescue training can be compared with fire programs that train new personnel to the Firefighter I level without conducting live burns of any magnitude. All the information can be provided; however, the student can't truly appreciate the lessons until he has experienced them firsthand. On the fireground, these include the lack of visibility, rising heat, and not knowing exactly where you are, let alone the location of the victim. Rope rescue has similar elements, height being only the most apparent of them. Obviously we can't train for everything, but we do need to train in context as much as possible. Those sterile techniques we learn in the classroom need to get dirty in the field to prove their worth. Training in context is a methodology by which more time is devoted to hands-on experience and less on lecture.

Some jurisdictions lack either a primary facility or adequate real-life sites. One way around this problem is to set up a platform, either on a tower ladder or similar apparatus. Setting up a tower ladder allows you to accomplish many tasks easily, anywhere in the district. You can use its rated points of attachment as anchor points. A tower ladder also allows you to change heights and locations in and around obstacles to suit your needs. This is no small advantage, and it relates to certain criteria that need to be observed prior to operating. The aerial has to be rated for the intended loads; it must be properly maintained and certified per NFPA standards. Also, the turntable must be manned continuously; a safety person must have an unimpeded view of the operation and be in control; all moves must be slow and deliberate, and the apparatus must be used within the manufacturer's recommendations.

The term *tower ladder* denotes an apparatus that has a platform or basket at the end of the ladder that can hold personnel. The use of straight-stick aerials concerns me—I prefer to use them in limited capacities. In my region of the country, there are a lot of older aerials in service that aren't rated for this kind of duty. They lack the tip capacities and jacking configurations needed for the evolutions being performed. This is where mutual aid companies can get together for a joint drill if your particular fire department doesn't have a tower ladder.

INCIDENT COMMAND

By now, virtually everyone in the fire service is familiar with the principles of incident management. The current terminology is changing from the incident command system (ICS) to the incident management system (IMS). Whatever its name, anyone involved in the fire service, law enforcement, emergency medical services, wilderness search and rescue, and host of other emergency-related fields must use some form of the management system to handle today's incidents effectively. For the purposes of this book, we will consider the fire department to be the parent organization, and the management system used will be the NFA model ICS. Several similar models are in use, differing mainly in terminology and title. It makes no difference which system you adopt as long as you use it consistently and all the people involved understand their roles within it.

For those departments that still don't employ an incident command system, here are some thoughts to ponder. SARA, the Superfund Amendments and Reauthorization Act of 1986, requires organizations that handle hazardous materials incidents to operate within a command system. NFPA 1500, *Standard on Fire Department Occupational Safety and Health Program,* Ch. 6-1.2, requires that all departments establish written

NFA Model Incident Command System

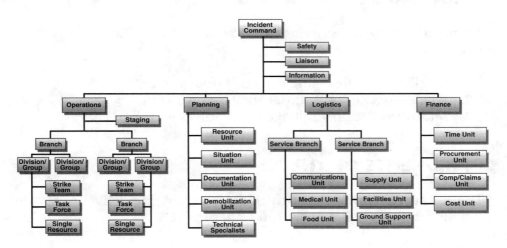

procedures for an ICS; that all members of the department be trained in and familiar with the system; that the system outline the responsibility for safety at all supervisory levels; that the system provide for personnel accountability at all levels; that the system outline safety requirements; and that the system provide sufficient supervisory personnel to control the position and function of all members operating on the scene. As you can see, if your homegrown system isn't providing at least these parameters, then you should adopt a nationally known system.

Why bother to discuss incident command in the context of rope rescue? I'm reviewing it because our operations can take place at the micro or company level with only a few members, or at the macro, multicompany level with many. Teams can operate independently of each other or in unison. In any case, you need to understand that you are part of a larger organizational structure and that you have to operate within its framework. The ICS provides procedures for controlling personnel, facilities, equipment, and communications.

Let's review some of the highlights of the ICS. The system begins from the time an incident occurs and lasts until such time as emergency services are completed. The commander is usually the ranking member; however, it can be anyone ranging from the first-arriving company officer to the chief. Passing off responsibility from officer to officer must be done in a coordinated way so as to accurately disseminate all the facts. Passing off command at the company level, however, can't occur more than twice. The organizational structure can be established and expanded as the incident dictates. Only qualified personnel should staff the positions, and they can come from agencies other than the fire service.

OPERATING REQUIREMENTS

A properly designed system can provide for the following types of operations:

1. Single jurisdiction/single-agency involvement.
2. Single jurisdiction/multiagency involvement.
3. Multijurisdiction/multiagency involvement.

COMPONENTS OF THE ICS

The following components working together interactively provide the basis for an effective operation under the ICS:

1. Common terminology.
2. Modular organizations.
3. Integrated communications.
4. Unified command structure.
5. Consolidated action plans.
6. Manageable span of control.
7. Designated incident facilities.
8. Comprehensive resource management.

ORGANIZATION AND OPERATIONS

The ICS has five major functional areas:

1. Command.
2. Operations.
3. Planning.
4. Logistics.
5. Finance.

COMMON ICS TERMINOLOGY

Branch: The organizational level having functional/geographic responsibility for major segments of incident operations. Organizationally, the branch level is between section and division/group.

Command: The act of directing, ordering, and/or controlling; also, the incident commander.

Division: The organizational level having responsibility for operations within a defined geographic area.

Finance: The entity responsible for all costs and financial aspects of an incident.

Group: The organizational level having responsibility for a specified functional assignment, such as ventilation, salvage, overhaul, or rescue.

Incident commander: The individual responsible for the management of all the incident operations.

Logistics: The entity responsible for providing facilities, services, and materials for the incident. They are the gofers of an operation, getting what you need.

Operations: The entity responsible for all tactical operations at an incident.

Planning: The entity responsible for the collection, evaluation, dissemination, and use of information about the development of an incident and the status of resources. This domain covers the technical specialists, including rope rescue personnel.

Single resource: An individual company or crew.

Staging: The location where personnel and equipment are assigned to wait on an immediately available status.

Strike team: Five of the same type of personnel, with common communications and a team leader.

Task force: A group of any type of personnel, not to exceed five members, with common communications and a leader, temporarily assigned to a specific mission.

What does all this mean? The bottom line is that the incident commander is responsible for everything on the scene of an emergency. Is it practical for one individual to handle all the various supervisory functions? No! That's why the ICS was developed, so as to delegate responsibility, dedicate resources efficiently, and mitigate incidents safely. Technically, if the IC doesn't delegate a function, it becomes his direct responsibility to see that it is accomplished. In the early stages of development, many times this is a necessary evil; however, as qualified personnel become available, they should be assigned to handle the various missions.

The ICS review process is to inform those who don't presently have a system to get one and those who do to review the terminology. Generally, when most departments respond to the initial call for a structure fire, they operate in a task force configuration. In larger districts, for example, this may mean three engines, two ladders, and a battalion chief. In rural situations, it may be two engines, two tankers, and a chief that respond to the scene. During this phase of organizational structure, we can assume that they are operating in the command section, dividing into branches.

Let's look at the larger department first. That three-engine, two-ladder assignment is going in, and additional information indicates that a large fire is spreading to another building. Depending on your run cards, an additional engine and ladder along with a rescue company are sent to answer the alarm. There are many ways to handle this situation, most of which are beyond the scope of this book. Suppose, however, that after these companies arrive and begin operating, a rope rescue becomes necessary.

You should attempt a rope rescue only after all other avenues have been explored and dismissed. The decision falls on the incident commander as part of the ICS process. The IC depends on all the other members operating on the fireground to be his eyes and ears. In this

ICS Structure for a Fire in a Multiple Dwelling With Exposure Problems

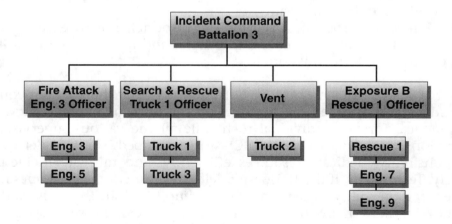

This is just one of the many ways to assign fireground tasks. Some departments that have above-average staffing assign (per SOPs) specific duties to incoming companies as they arrive. The companies break up into teams that address specific areas of the building. Example: The first-due truck searches the fire floor, the second-due truck searches the floor above, while others vent and perform forcible entry and search.

case, let's say that the roof man and the outside vent man have spotted two civilians hanging from a fourth-floor window. They relay this information to the incident commander, and he quickly determines that the best course of action is a dangerous but calculated rope operation. Depending on many factors, such as staff levels on apparatus and the percentage of fire involvement, the entire operational focus can change to support the rescue effort. Sufficient personnel will be assigned to undertake the mission; suppression strategy might change from an aggressive offense in one part of the structure to a hold-the-line defense in another. Basically, the fireground focus now is on protecting the members who are hanging from the rope, as well as the victims they are going to rescue. This strategy follows the premise that life comes before property and that buildings are replaceable but people are not. Although the rope rescue contingent itself may be small, clearly the rescuers are operating within a larger organization readily adapting itself to the changing context of the incident.

Once those personnel get to the roof and begin their rescue, a smaller but equally important substructure comes together. This is another area where a number of options are available, depending on local practices. Let's look at six truckies who converge on the roof and set up for rope rescue. Generally an officer, if not a senior firefighter, makes it to the roof and gives the go/no-go decision. The first and second firefighters should be in harnesses ready to go over the edge to handle the two victims. The third firefighter sets up the anchor points. The fourth assists with anchors and rappel lines. The fifth is the edge man, who maintains eye or voice

ICS Structure to Support the Rescue

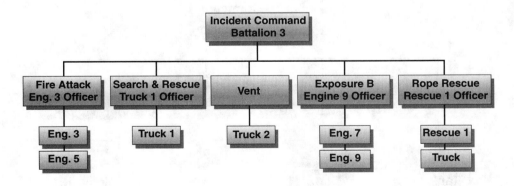

Additional alarms are initiated as the fire magnifies in scale and the scope of the operation grows. Incident commanders must be aware of the ramifications when switching from one type to another.

contact with the victims while the evolution is being set up. The officer or senior firefighter can act as an overall safety person, stepping back to assess the overall picture. There's not much time, and they must set up rapidly. To some, this scenario may be utopian in terms of the personnel available; to others it will be the norm. These are issues that must be discussed within your own department and settled prior to any operations.

Applying the acronym LAST to this past evolution, the locating was done by two firefighters. Access was made from the roof, which provided the operating platform. The victims were stabilized by the edge man, as well as by firefighters on the ground giving verbal reassurances. The transport was the rappel down the side of the building to the ground.

This tabletop scenario can occur anywhere; however, the emphasis here is on the dynamic change of the strategy and tactics. The IC envisions one plan based on his size-up of the prevailing conditions, but the fire has other intentions. Once the rescue effort is underway, the commander has limited staffing to handle both buildings. Reviewing the ICS matrix, you'll see that this IC activated the command function early and was headed toward other operations as additional assignments arrived.

In the next scenario, imagine that a small aircraft has crashed into a TV tower. The tower is located in a suburban and rural jurisdiction. This

ICS Structure for Initial Crash Investigation

incident starts with a call to the county fire alarm to put the volunteer fire department on standby. The call originated from the local airport, which is receiving an ELT (emergency locating transmitter) signal from an overdue aircraft. The airport controllers have tried several times to contact the pilot by radio but to no avail. The fire chief waits until his station is manned and he receives an update. The controllers get a preliminary bearing on the ELT signal, and a ground search commences with fire apparatus and police units driving through the more rural areas first, principally with the rationale that if the aircraft had gone down in a populated area, someone would have seen and reported it. The airport is trying to get portable radio equipment together that can triangulate the signal more precisely. Suddenly a fire alarm indicates that Channel 5 has just lost its transmitter, which just so happens to be in line with the initial ELT bearing. At once the chief assigns apparatus to that location. On his arrival, the chief finds a small fixed-wing aircraft lodged in the upper portions of the transmitter tower. The tower appears to be relatively undamaged and in no danger of collapse. Sizing up the environmental factors, the chief sees that nightfall is fast approaching and that the temperatures will drop to around 30°F.

Prior to his arrival, the chief had envisioned a plane crash on the ground. In the worst case, it would have been on fire with the occupants trapped. He now establishes command and starts assembling the

ICS Structure for the Tower Rescue

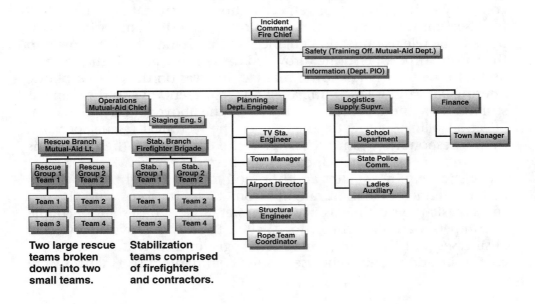

rescue organization. He has to expand his ICS from the more routine approaches because this sort of rescue will be far more difficult. He foresees this to be a single-jurisdiction/multiagency response because his department simply isn't trained or equipped for this type of operation. While his on-scene personnel stabilize the situation as best they can, numerous calls are made for help from other organizations.

This ICS was set up by the chief, and here's how and why he did it:

1. The IC was the chief of the department.
2. The safety officer was the training officer from the mutual aid department.
3. Liaison was performed by the IC himself.
4. The information officer was the department's public information officer.
5. The staging officer was a lieutenant from Engine 5.
6. The operations officer was an assistant chief from a nearby city.
7. The planning officer was a firefighter who worked as an engineer.
8. The logistics officer was the department supply supervisor.
9. The finance officer was the town manager.
10. The rescue branch officer was from a mutual aid department.
11. The stablization branch officer was a firefighter in the construction field.
12. The rope rescue teams came from two nearby fire departments, as well as a mountain search and rescue group.

Additionally, each rescue group was supervised by a team leader or company officer from their respective home organizations. The planning section was comprised of technical specialists from various disciplines. The television station engineer was on hand to isolate power to the tower. Airport personnel provided data on the aircraft that proved useful in the rescue. The tower manfacturer was on the cellular phone offering data and was sending a regional representative to the scene. A structural engineer conferred with all of the above and devised a stablization plan to allow for the rescue. The technical rescue specialist from a responding team orchestrated the details with the operations chief and the rescue branch officer to effect the rescue.

Logistics required shelter and hot food for the firefighters. It was to be an extended operation ranging from 12 to 18 hours, and the rescuers' basic physical needs had to be met. School buses and a couple of campers were brought in for shelter. Food was provided by local businesses and the Ladies Auxiliary. Logistics also got a state police communications unit to be used as the command post. Finances were tracked by the town manager for future reimbursements. The insurance companies representing the aircraft owner, a renter, were on the scene, as were those responsible for the TV station's antenna.

This has been a quick look at the macro level of an incident using the ICS. The vast majority of any department's calls can and will be handled by a smaller structure.

STUDY QUESTIONS

1. Which ICS position is responsible for tactical operations at an incident?

2. The overall goal that the incident commander wants to accomplish is known as _____.

3. Name the four components of the search-and-rescue acronym LAST.

4. Organizations that handle hazardous materials incidents are required to use the ICS by what act?

5. Name the five sectors of the incident command system.

Chapter Three

Safety and Personal Protective Equipment

S afety is one of the most important subjects in any book of this type, if not the most important. All too often, it is relegated to the end of the text, near the credits or glossary. I feel that safety should begin this sort of book. Keeping a semblance of order suggests that you must first assess and understand your basic need for safety before dealing with the actual hazards themselves. Once you recognize that need, you'll be in a better position to truly understand what you're getting your personnel into. You have to look past all the hype, fancy equipment, and fads. When you complete your homework, you'll discover that rope rescue is really a safe and time-tested discipline, albeit a demanding one. Still, behind all the best preparation, an element of danger lurks, and individuals have been injured and killed performing these techniques. One or more of these factors can contribute to any mishap:

1. The human component.
2. The equipment component.
3. An environmental upheaval.
4. Being in the wrong place at the wrong time (bad luck).

This matrix takes the four primary factors and incorporates them into a systems approach to rope rescue safety. A systems approach, whereby you view components as interdependent and unified, is essential. A firefighter's cognitive skills are part of the system and not to be overlooked. Too often you focus strictly on the physical and don't build the human factor or the effective attitude into the operational system. You become totally absorbed with the psychomotor skills (raise the ground ladder) to the exclusion of cognitive ones (avoid the overhead obstacle, don't trip, don't get wrapped in the halyard).

The fire service, for all fireground operations but particularly high-risk rescue, cannot permit the cognitive skills or affective attitude to be one of the weak links in its systems. There is simply too much at stake.

Let's examine the four primary categories in rope rescue system safety, with an awareness of the importance of the human, as well as the physical, components of the system.

THE HUMAN COMPONENT

In the training section in Chapter Two, we looked at the definition of training itself. Training closes the gap between where we are and where we want to be in our knowledge and competence of a given skill. It sounds simple enough, but how do we achieve it? Three phrases from the profession of education separate the domains of learning: the cognitive, the psychomotor, and the affective. The cognitive domain pertains to objectives in which the firefighter needs to recall intellectual knowledge or concepts. The psychomotor domain deals with objectives in which the firefighter must demonstrate a physical proficiency. The affective domain involves the area of appreciation or attitude that the firefighter must have to accomplish his goal. One of the predominant methods to ensure that the firefighter has grasped this triad of domains is to test them. A cognitive test, for example, could be a written exam on equipment specifications—straight, factual information. A psychomotor test could be a hands-on, step-by-step evolution of an assigned task. The affective domain, in my experience, is the hardest to evaluate with firefighters. You can quantify their knowledge of a subject or their ability to perform a job, but how do you put a numerical score on something as vague as attitude? Yet it's attitude that is the basis of safe operations, no matter what the job is. A positive attitude goes a long way. It's the sound frame of mind that best judges a situation and makes a good decision. It's the positive attitude that takes the high road and balances the risk-to-benefit ratio of every incident. The best way to score attitude is through exercises whereby you examine the thought process of how one arrives at a conclusion. Another method is a long-term observation of the firefighter's work habits to assess his liabilities. Judge the whole individual rather than pieces. In essence, there are three components to human performance, and if any one of those components isn't up to par, any one of us can have a bad day. Many other psychological factors are involved, of course, and this in no way attempts to limit them; rather, they are offered to show how important training is to safety.

Essential for both the individual and the overall team is a collective mindset to aggressively learn, master, and maintain rope rescue skills. All members must understand the hazardous nature of rope techniques and be willing participants. We all have our different fears or appreciations of the dangers involved. Members must constantly monitor each other for subtle changes in their normal mannerisms—changes that might be critically detrimental to their performance. Regularly I watch students who forget where they are in relation to the ground. They lose sight of their surroundings and, were it not for the safety measures in place, might be severely injured. The differences between engine and ladder companies provide a good analogy to depict the variant levels of danger. In general terms, if an engine man were to trip over a hoseline, he'd probably hit the floor but remain relatively unscathed. If a truckie on a rooftop were to trip over an obstruction, however, he might take a ride off the roof or down a shaft. I use these to illustrate the need to be cognizant of your environment at all times. While many texts have been written on training, safety, and psychology, this book will treat such factors as fundamental assumptions. The human component is a multifaceted variable encompassing far more than rope rescue skills. We'll take a basic approach to safety by using the domains of learning as the foundation of safety.

Let's consider the components of a behavioral objective. Many lesson plans formally take into account the components of the learning process, commonly using a format of A, B, C, and D:

A = Audience, or who it is intended for.
B = Behavior, or what the required performance is.
C = Condition, or what tools and equipment are required.
D = Degree, or what the accuracy and time constraints are.

While lesson plans also list the objective as either cognitive or psychomotor, many of our activities aren't covered by lesson plans and so need an umbrella principle instead:

Cognitive domain = knowledge.
Psychomotor domain = tasks.
Affective domain = attitude.

Consider the following examples. You issue new rechargeable flashlights to your members. They look like the old units, so members maintain them the same way. After a short period of time, the units begin to develop operating problems. One by one, the lights are turned in as defective. Eventually you discover that when the units run low, the members are purposely draining them until they are dead. These new-style lights can be charged as often as is necessary without running them down. Is this a safety problem? Of course it is! Apply the domains to this situation this way:

Knowledge: Require that the members read and follow the manufacturer's information.

Tasks: Require the members to operate and maintain the lights as directed.

Attitude: If you don't operate and maintain a unit as required, it may fail at a critical time. Your life may depend on it!

Here's another example. You assign members to set up a lowering system on a steel structure. They grab the equipment bags and ascend to the top. Once they get going, they realize that they're a little short on hardware and slings. Since this is a training session and not a real emergency, they opt to wing it with a bare minimum of equipment. Afterward you debrief them and point out their shortcomings. Using the domain approach, members should do better in the future.

Knowledge: The members need to know what equipment is required to set up the system. This knowledge comes from specific training and exercises.

Tasks: The members must pack sufficient equipment and transport it to the operating site. If additional equipment is needed, they must either get it themselves or have it delivered by other members.

Attitude: They must understand that operating with bare minimums and jerry-rigging aren't ordinarily acceptable.

EQUIPMENT COMPONENT

Hazard abatement is a process whereby hazards are reduced by way of a planned program. In industry, hazards are minimized by engineering safety systems into the design, having machines replace humans, permitting minimal exposures, and providing personal safety equipment. While the first three methods may work in general business environments, we as firefighters knowingly enter dangerous situations. Ultimately, no book can tell you what is a go and what isn't. Every situation is different, even if it occurs in the same location twice in the same day.

When we perform our jobs, we need to protect ourselves. One of the primary methods is with personal safety equipment, and the foremost area to protect is your head. As we all know, even a minor head trauma can lead to a debilitating injury. One firefighter I know of was operating at a brush fire in an urban area of the city. He trotted back toward the engine for more hose, when suddenly he struck his head, which was properly helmeted, on a deck overhang. He was taken to the hospital, examined, released, and was placed on the injured list. He suffered from seizures, dizziness, headaches, and general weakness—and could have been still worse if he hadn't been wearing his helmet at all. He slowly recuperated; however, it was a stressful, frustrating process for him and those who love him, which brings me to another point: If you don't take safeguards for your own sake, *do so for your family's sake.*

The type of head protection you wear should be approved for the environment in which you plan to operate. The fire service uses the NFPA standards as a basis when purchasing certain personal protective equipment. NFPA 1972, *Helmets for Structural Fire Fighting,* covers performance criteria for fire helmets.

Protect yourself for the highest level of hazard expected. This means that if you are going to perform a rescue on or in a burning building, you must wear a fire helmet. Other types and styles of helmets can be worn while performing rope rescue. When operating in a confined space, some prefer to wear a low-profile helmet—the size and overhang of a conventional fire helmet may cause it to get hung up and create problems. Different helmets are available because different disciplines have created them to suit their needs.

Not only do helmets provide head protection, they can protect the eyes as well. Helmets should be properly fitted and maintained, and you should always use the chin strap.

The next area of concern is the torso. In general terms, you should wear clothing that will protect you from the environment in which you are operating. A wilderness rope rescue may find you wearing a wool shirt and dungarees covered by a Gore-Tex® rain suit. That provides you with warmth, flexibility, and waterproofing. A fireground rope rescue requires you to don a full ensemble, known either as bunker gear or turnout gear. Again, look to the NFPA standards for assistance. NFPA 1971, *Protective Ensemble for Structural Fire Fighting,* lists the performance criteria for the garments we wear while working in hostile environments. We must protect ourselves while operating in or close to the fire scene and in the event that the rope rescue doesn't go as planned.

There are other environments in which we perform rope rescues, such as on ice or in water. Generally speaking, when we operate on ice-

Left to right: cold-water immersion suit, flotation coverall, Type 1 life jacket.

covered lakes, ponds, and rivers, we wear full-body flotation gear, otherwise known as the gumby suit. This closed-cell, neoprene suit totally encapsulates the body, hands, feet, and head area. Barring any leaks,
it provides both flotation in the water and insulation from the cold. It
got its nickname from the cartoon character Gumby, since you must
shuffle along to walk in it.

All too often, you see pictures of firefighters attired in full bunker
gear working on ice. This poses a great danger if the individual happens to fall through and starts submerging. While it's true that bunker
gear will float in water, it will only do so for a short period of time.
When you are going to work on or near ice, you must wear the proper protective equipment. One particular benefit of having such a suit
is that it can be used for water-in-the-basement calls. The firefighter
will never get wet, and wearing it is a great public relations tool.
Remember to secure all power sources prior to entering the water.

A variation of this is the coverall-type flotation suit. This suit is
designed to be worn over your street clothes, but it can also be used for
day-to-day activities where a person might fall into water. It doesn't
protect your feet or hands the way the gumby suit does; therefore, it
doesn't afford the same level or length of protection.

These first two types of suits are designed to provide flotation, so you won't have to wear a separate life jacket. Whenever you enter or fall into water, you must wear some sort of flotation. Most fire departments use a commercially available U.S. Coast Guard-approved device. For general purposes, any of these approved devices are usually sufficient. For WETEAMS (water extrication teams) or any other organization that specializes in water rescue, more specialized apparel is available.

Your feet are the next area of concern. Different types of environments require different types of protective footwear. Appropriately selected footwear is important to everyone performing rope rescue skills. The shoe or boot should be comfortable yet still provide comfort, ankle support, traction, and repellency as a minimum. Thermal protection, waterproofing, and puncture-resistant characteristics are other desired features for certain conditions.

Firefighters operating on the fireground are limited to a small selection of boot choices. An additional part of their protective ensemble includes footwear as specified in NFPA 1974, *Protective Footwear for Structural Fire Fighting.* Over the years, fire boots have changed for the better. The ever-popular three-quarter boot is giving way to the shorter bunker boot. Many firefighters are donning bunker pants; therefore, they don't need the additional weight of the rolled-down three-quarter boot. In the past, some firefighters found the rubber boots to lack sufficient ankle support. The newest rubber boots on the market have been improved to give their wearers that "boot fit."

Leather boots comprise the second type of approved footwear for firefighting. These boots meet the standard's criteria and are lighter in weight, better fitting, and more comfortable by general consensus. Some firefighters maintain that once the leather is scraped, however, particularly near the toe, a boot loses its waterproofing features. Before purchasing boots in large numbers, field test each type from several manufacturers using your toughest people.

Unfortunately, I still see firefighters wearing nonapproved or substandard boots while performing structural fire attack.

Boots or shoes for other areas where rope rescue is performed are too numerous to list. If you are operating in an industrial area, the fire boot or approved work boot may suffice. If you're operating in a rural or wilderness area, then a hiking or climbing boot from an outdoor supplier may suit your needs. If you're operating in water or doing flood rescue, then a soft-soled boot, sneaker-type shoe, or kyacking shoe from an outdoor supplier may do the trick.

Your hands are the next area of concern. This topic draws attention from many in the rope field.

The conventional sport climber generally wears a light leather glove with the fingertips cut off. This gives his hands some protection plus good fingertip dexterity for finding handholds. Some wear no gloves at all or, if they do, take them off for rappelling. This prevents them from sliding down the rope too rapidly, possibly burning or otherwise damaging it.

On the fire service side of the issue, NFPA 1973, *Gloves for Structural Fire Fighting,* covers the performance criteria for gloves worn by firefighters while working a structural fire. Some firefighters don't want to do rope training evolutions with fire gloves on because contaminants make the rope slippery. We all know that it's important to keep a rope as clean as possible; however, to train for fireground operations, I feel that personnel should get used to handling rope while encumbered by such gloves, which are bulky and have limited fingertip dexterity. This isn't to say that members can't wear a clean pair or have gloves set aside for rope purposes. Fundamentally, we need protection from the environment and mechanical injury in the broadest sense. I believe in wearing gloves anytime you climb around man-made structures. We in the fire service should train with gloves because we need to protect ourselves from the fire environment. During a roof rescue or emergency egress, you're not going to take the gloves off, so you'd better get comfortable with them now. In my area of the country, the winters are long and the wind chills go way below zero. There aren't too many arguments as to whether gloves should be worn or not. While operating, usually you do everything you can while wearing them, then finish up the detailed tasks as fast as you can when you take them off.

The lack of adequate lighting for the work area is the next area of concern. Providing proper lighting is a constant battle against a variety of problems, including insufficient candlepower, dead or dying batteries, environmental factors, incorrect lenses, and mechanical damage. Personal lighting equipment entails handheld or helmet-mounted lamps, not apparatus or hard-wired lights. Insufficient illumination of dark or smoke-filled areas has caused injuries to firefighters and slowed operations.

For years, the fire service has used rubber bands to attach disposable lights to leather-type fire helmets. This provides a small light with one drawback: You must move your head to shift the beam. The next helmet-mounted light to show up in large numbers was the Mini Mag® light attached to the composite fire helmet. This light is adjustable from flood to spot, and the batteries are replaceable; however, you still have to move your head to aim the beam. Several small flashlights on the market allow you to pivot the light to where you want it to be. In the old days, coal miners wore carbide lamps on their helmets. Today, flashlights mounted on helmets can be powered by smaller belt-mounted rechargeable batteries.

The other popular form of personal lighting is the handheld or clip-pable flashlight. For years, truckies have slung lights around their shoulders. Many units are available that run on rechargeable or replaceable alkaline batteries.

Some issues you must consider are how the environnment will affect the duration of the power available, what the operational duration of the light is under good conditions, the numbers of lights available, and how many are available as backup. The other question that comes up frequently has to do with provisions for explosive atmospheres. Although some units are rated for such environments, once they've been damaged or even repaired, there's no guarantee that they'll still be explosionproof. Firefighters are notoriously brutal on their equipment, and more than one repair technician has advised me never to trust electrical items to be entirely safe. This goes for almost anything from portable radios to flashlights to monitoring devices. It doesn't mean that you should simply keep specific equipment out of the hands of those who may abuse it, however.

To wrap up lighting, think about your firefighting experiences and your day-to-day requirements. Many of my department members use a highly popular rechargeable flashlight that clips to their bunker coat. Even a structure fire of moderate size, in warm weather, and of short duration creates a traffic jam at the chargers back at the station. The other problem is that many of the rechargeables are sealed, and new batteries can't be installed as they could in the past.

HARNESSES

Some departments consider rescue harnesses to be a part of safety equipment; others as team equipment. Either way, you need them to perform rope rescue. NFPA 1983, *Fire Service Life Safety Rope and System Components*, Ch. 4-3, lists three classes of harnesses. In their essential features, they are:

Class I— That which fastens around the waist and thighs, and under the buttocks, and is designed for one-person loads only.

Class II— That which fastens around the waist and thighs, and under the buttocks, and is designed for two-person loads.

Class III— That which fastens around the waist, thighs, and under the buttocks, and over the shoulders, and is designed for two-person loads and whenever inverting can occur. This harness is mandatory for confined space rescue because it allows you to pull someone through narrow spaces and remain upright.

Years ago, going down a rope was called sliding for life; today it's called descending. In the days of hemp, firefighters used pompier or ladder belts as harnesses. The major problem with these is that they tended to transfer all the body's weight to the lumbar region of the spine. This was uncomfortable, leading to injury for some and short working times for all. Eventually members turned to sport climbing equipment as an alternative.

The harness standard didn't come out until 1985; however, sport equipment and techniques were already taking hold in the field by then. The inherent problem with sport equipment is that it tends to be lightweight and thus unsuitable for the fire department. Having said this, let me qualify that statement. Many of the currently available sport harnesses aren't designed to handle the two-man loads that rescue may demand. Rescue-grade harnesses have to be constructed with 6,000-lb. webbing as a minimum. In the international arena, particularly in Europe, the philosophy follows a different approach, whereby harnesses are made lighter in weight primarily because any stress over 12 kN (2697.6 lbs., one kilonewton equaling 224.8 lbs.) can cause serious consequences. Constructing a harness to handle well over 12 kN isn't prudent because of the effects of high g forces on the human body; also because such decelerations will cause some other component in the system to fail anyway.

Between the time of the ladder belt and sport equipment, members used webbing to tie harnesses. This technique was derived from the military background common to many firefighters. The ever-popular Swiss seat showed up everywhere, and it was tied with both rope and webbing. If you have ever used this device, you know that it's uncomfortable to be suspended in one for long periods. Many beginner training programs use them for economic reasons, and more experienced rescuers use them for quick or one-person emergency operations only. This category encompasses many of the similar quick-tie devices used in the field.

During this time we also went from three-strand rope to one-inch tubular webbing for comfort and strength; however, it still didn't entirely meet the need. Individuals tried two-inch webbing for peace of mind, but it wasn't very comfortable. I purchased my first compliant harness around 1985 and was quite happy with it. You stepped into each leg loop and connected the waist strap and you were set.

Today, harnesses come with a variety of options. Do your homework and ask around. It's like shopping for a car—to each his own.

COMMUNICATIONS

One of my pet peeves is good radio communication. Many of my colleagues consider the portable radio to be a life safety device and critical to incident management. Every team, group, or company must have at least one portable radio during an operation. More and more departments are issuing portables for every riding position so that their members can be more flexible and the span of control can be maintained. Portable radios are available in a wide variety of designs, wattages, capabilities, sizes, and prices.

Communication between companies or individual firefighters can be more difficult than imagined. Many factors can impede effective communications, including weak batteries, topography, structural materials, natural interference, and operating on the wrong channel. Each member should begin his shift with a freshly charged battery. Many of the batteries that are currently available can last twelve or more hours while monitoring and if few transmissions are made. Some firefighters will use a portable all day and at the end of the shift put it in the apparatus charger. What they don't realize is that many of those units are trickle chargers, designed to maintain the current energy level without restoring it to max levels. Do this several times in a week and soon you'll have a dead battery. Among other things, this is a safety issue. Many departments have a primary alert frequency for apparatus movement; however, once on the scene, they switch to a fireground or rescue frequency. Complex incidents may require several tactical channels for proper control to be maintained. Two results can occur from improper or insufficient frequencies: First, if intrasite communications are on your primary frequency, members in trouble might never be heard. Second, companies not aware of frequency changes can't get their messages to Command if trouble arises.

Topography and troublesome construction features can also dramatically impede good communications. Public safety agencies use a wide variety of frequencies in the various portions of the radio spectrum. Frequencies are grouped and allocated by the FCC. We'll call them groups for simplicity and list them as low, mid, high, and ultrahigh frequencies. Each group has different operating characteristics and is chosen by professional radio contractors for your particular needs. The wattage output on a portable radio can be adjusted from, say, one to five watts at the push of a button, a lower output allowing you to run longer. Operating in and around steel or concrete buildings can hinder the radio signal. Many fire service portables are midrange 150 to 175 megahertz units, and there are times when a five-watt signal can't reach companies one floor below in a city hospital or hit a repeater on an adjacent hill. Companies that regularly operate in these types of structures, aboard ships, or in caves should carry high- or ultrahigh-frequency portables, since they seem to provide more reliable communications. When performing rope rescue in a dense urban environment, in a confined space, or below ground, carrying a portable slung over your shoulder can get in the way. Worse, it might be totally ineffective. Instead, members can carry them inside their clothing, use a chest pouch, or attach them to the harness and use a remote mike.

When working in explosive atmospheres, the same considerations apply as were mentioned in the section on flashlights. Ask yourself why you are in that environment and to what degree the portable is explosionproof. New technologies are constantly being developed to deal with a host of radio deficiencies.

Hardwired or confined-space intercom systems are now available to facilitate smoother and more recognizable communications. The major drawback to hardwired equipment is that, when exiting a con-

A hardwired confined-space communications set.

fined space, you must leave by the same route you entered. Many of these systems are intrinsically safe, even in hazardous atmospheres.

Intersite communication involves talking between the incident scene and dispatch. The usual methods include portable to base, mobile to base, and portable to apparatus repeater and then to base. There are times that, for whatever reason, the usual methods are ineffective or inoperative. Accordingly, there are a few ways to handle such a situation. If possible, use a standard landline telephone to get through to dispatch. The most popular method of alternative communication is the cellular telephone. These systems tend to have greater coverage than municipal radio systems. Telephone companies are in business to provide a service to their customers, and companies with the best coverage tend to make the most money. The trick for you is to be in range of a cell, which is the receiver for your signal. From there you have as much access to the world as you would from your home telephone. This is helpful in rural or wilderness areas, particularly if the voter or repeater is inoperative. Another option involves using the county or state police radio equipment. Many of these departments have excellent radio systems in place, particularly the state agencies. Using their equipment will generally suffice.

A final option worth mentioning is to rely on the public utilities. These companies tend to have radio systems in place to provide coverage across their service areas, which can be vast. One statewide power company near me has radio coverage that rivals that of the state police. Most of us take radio communication for granted simply because it works day in and day out. Its frangible nature can often be overlooked in the general scheme of things.

EQUIPMENT FAILURE

On the other side of the equipment issue is the possibility of failure. Without cutting too much into future chapters, let's look at a few situations that can get us into trouble.

First, it helps to divide our equipment into two categories: hardware and fabric. Hardware includes carabiners, descending devices, ascending devices, and other auxiliary equipment. Safety issues involving hardware usually center around correct use and loading. These devices are designed to function in a specific plane or direction and, when stressed inappropriately, distortion or failure can result. A common situation is a cross-loaded carabiner on a system or in a descending evolution. Instead of the forces pulling along the major axis, they pull along the minor axis, for which the unit wasn't designed.

Fabric refers to the webbing and rope. Webbing is considered disposable and takes the brunt of abrasive surfaces so as to save the rope. This doesn't mean to use it a few times and throw it out, however—simply that it is more cost-effective to abrade a couple of 10-foot webb slings than to damage a 50-foot section of rescue rope. Rope is the main load-carrying medium throughout the length of any system.

The primary enemies of webbing and rope are abrasive and sharp edges. To minimize fabric failure, use edge protection or remove the trouble spots. Chapter Four will address this subject in more detail.

In any system failure, it's usually one of the weaker links that starts a chain of undesirable effects. Look at the top of the system, at the anchor point. Is there more than one? If so, what type are they? Is there a backup system? How much and what type of fabric is used? Access and transportation systems (i.e., edge protection, rope grabs, ropes, pulleys, racks, eights, carabiners) typically fail because (1) dynamic loads and unforeseen stresses are placed on the system, (2) the equipment is used incorrectly, (3) components have been jerry-rigged in a pinch, (4) too much mechanical advantage has been created, or (5) too many people are pulling. It is axiomatic that a simpler system is less likely to fail because there are fewer links in the chain.

ENVIRONMENTAL COMPONENT

In the section on personal safety equipment, I listed some basic items that should be used, depending on circumstances. In terms of rope rescue, some common environments require an additional degree of planning. Again, rope techniques don't stand alone—you must combine them with all of your other firefighter skills and experience to carry out the mission and return intact. Generally speaking, rope evolutions are relatively safe; however, the environment in which we operate always plays a role and often is the cause of failure. As firefighters, we must never be so naive as to think that as long as a system is right, nothing can happen.

Consider the deadliest environment of them all, the structure fire. Some of its inherent dangers include the lack of oxygen, toxic gases, high heat, and smoke. If you're going to perform a rope rescue in and around the fireground, you're going to need respiratory protection—an approved positive-pressure self-contained breathing apparatus, or SCBA. Firefighters must be proficient at setting up and performing their missions with their masks on. The mask shouldn't make too much of a difference. For some, however, it's a distraction that must be overcome. Many brands are available, but whatever mask you choose must comply with NFPA 1981, *Open-Circuit Self-Contained Breathing Apparatus for the Fire Service.*

Granted that the problems of setting up a rope system on or in a burning building are too numerous to mention, I'll list some of the major ones:

1. Finding suitable anchor points.
2. Maintaining visual and verbal contact with team members.
3. Rapidly changing conditions.
4. Sharp edges such as glass, flashing, and gutters that can cut your rope or webbing.
5. Heat and flames impinging on the rope.
6. The area of refuge being overtaken by fire.

Confined spaces present another dangerous environment, claiming many lives and injuring many more. The subject of confined space rescue is presently in vogue, in part due to 29 CFR 1910.146, *Permit Required Confined Spaces,* which has had a tremendous impact on private industry and the fire service. This type of environment is noteworthy because of the number of rescuers who fall victim to it.

A myriad of hazards are found in confined spaces; however, respiratory protection could eliminate many of them. Initial and continuous air quality monitoring, coupled with appropriate respiratory protection, is key to success.

Once a firefighter enters the space, he must perform all of his activities with his mask on. Some can overcome this distraction with practice. Using respiratory protection in such environs can prove difficult depending on the size of the space. Would-be rescuers have been killed by removing their SCBAs to squeeze through an opening, only to have their masks inadvertently pulled from their faces.

Currently, many industrial brigades and municipal fire departments are using supplied air respirators for those entering spaces that would preclude having an SCBA strapped to your back. This respiratory device supplies air through a hoseline, either from cascade bottles, a specialized air compressor, or a series of smaller air cylinders from an air cart stationed outside the void. The wearer has virtually unlimited air. In the event of a failure of any kind, he has a five- to ten-minute bottle to rely on for his escape.

Consult CFR 1910.134, *Respiratory Equipment,* for more details on such devices.

The great outdoors is truly the ultimate environment. Here we must endure heat, cold, wind, snow, rain, fire, flooding, and whatever else Mother Nature throws our way. Any of these factors can cause injury or death to a would-be rescuer. Extremes of temperature can rob us of our senses and judgment. Rain and snow can make conditions slippery and blinding. High winds can push us off mountaintops, bridges, and towers. Clearly, repetitive evolutions and timely use of rope techniques can save your life.

Industrial environments are ominous because they are both exotic and man-made. Although many hazards can be described, the one I want to discuss is noise. Often enough, rope rescue has to be performed around operating equipment. Pump stations are prime examples of where you may be able to isolate some of the machinery; however, it is nearly impossible to shut it all down for obvious reasons.

Even brief exposure to loud noise without ear protection can prove deafening, hampering your ability to perform. Noise also makes communications difficult, if not impossible. Rescuers operating in high-decibel contexts must use suitable ear protection. Traditional ear muffs get in the way of helmets or when working in tight spaces. A variety of squeezable foam or low-profile ear sets are available and should be used.

The last adverse environment is a matter of universal threat: being in the wrong place at the wrong time. Why this last category? Because there are those among us who sometimes forget how dangerous our jobs really are. Even if you train extensively, equip yourself properly, control the incident, and keep known risks to a minimum, you can become a casualty. For all our strides in reducing death and injuries, we cannot forget the imminent danger of our profession. It doesn't matter whether you belong to a 50-call-a-year company or a 5,000-call-a-year company—members in either organization can fall prey to circumstances beyond anyone's control. The best you can do is to pay attention to those things that are within your power, keep your eyes open, and always give yourself an out.

STUDY QUESTIONS

1. According to education theory, a hands-on, step-by-step evolution is used to test ability under which domain of learning?

2. What class of harnesses is recommended if inverting is expected in a rope evolution?

3. The two categories of rope rescue equipment are _____ and _____.

4. What are the primary enemies of rope and webbing at an operating site?

5. The supplied air respirator is gaining popularity in protecting firefighters in which type of rescue environment?

Chapter Four

Rope

The first three chapters cover some of the important administrative concepts that you must understand prior to performing any rope evolutions. The principle of training in context is currently in vogue, meaning more time is spent gaining hands-on experience and less on lecture. Even before this method became popular, however, many students failed to grasp those boring classroom, skull-busting, cognitive lessons. Studying nomenclature and technical material was for officers and training guys, not those actively working in the field. It's this lack of concern for the details of any firefighting topic that can spell the difference between success and tragedy.

By analogy, consider a recent conference of instructors and line officers. The question came up: What kind of threads are on your 1¾-inch hoselines? In an area dominated by iron pipe threads, and with your department adhering to national standards, what do you think could happen on a mutual aid call, hooking up your hose to another department's pump? Pay attention to detail, technical or otherwise!

This chapter is devoted to webbing, rope, and other cordage. You need to understand some of the nonglamorous details about this fabric. Taking it for granted can quickly turn you into a statistic.

HISTORY

We'll consider the modern fire service to have been born in the era just after World War II. As in most other fields, this period has seen the all-time greatest growth in technologies and skills.

The rope fibers predominantly available in the early days were the natural-fibered manila or hemp products. These natural-fibered three-strand ropes were the primary weight-carrying medium well into the

1970s. What's scary is that you can still find them being used for rescue purposes today. Old chiefs have old ways. Prior to the changeover to synthetic fibers, the two natural mainstays were the 5/8-inch rope, which had a breaking strength of 4,400 lbs. new, and the 3/4-inch rope, which had a breaking strength of 5,400 lbs. new. That 3/4-inch rope was quite a beast, but modern fabrics and techniques can construct a rope of the same strength at a fraction of the size. Among sport climbers, synthetic-fibered ropes had already been in use for some time. Being dynamic in nature, they were generally considered inappropriate for the fire service. Synthetic materials such as nylon were invented before World War II but didn't gain acceptance into the mainstream fire service until the 1980s.

So as not to take it for granted, what exactly *is* rope, anyway? Rope is a highly versatile rescue tool. It is low in cost and requires little maintenance but great care. Without it you could be stranded high above or well below the rest of the world. Virtually every fire apparatus in existence has some on board for one reason or another. Rope can be divided into two broad categories: that for utility use and that for life safety. The thrust of this chapter will be on rope's life safety use, since that is where our futures are most at stake.

Whenever I teach this subject, I generally start with NFPA 1983, which gives the criteria for the components we use. The logical sequence is then to cover materials or fibers, construction, and management of rope. The primary reason I reference the NFPA standard is to be comprehensive. I don't pretend for one minute that this or any other NFPA standard is beloved by all in the fire service. I also realize that no standard is a shield that will protect us from the dangers we encounter on the job. What NFPA 1983 does provide is criteria for design, construction, and performance of life safety components used by firefighters in critical operations. It doesn't cover any specific techniques; rather, it gives you a baseline for physical components to perform rope rescue evolutions safely.

The original standard came out in 1985 and was entitled *Fire Service Life Safety Rope, Harness, and Hardware.* In 1990, it was revised. When revised again in 1995, it was given its present title, *Fire Service Life Safety Rope and System Components.*

Many firefighters out there still aren't fully aware of the standard or its impact, so I'm taking this opportunity to promulgate it. If this educational forum saves one life, then it's worth the time. Every organization should have at least one copy of NFPA 1983 available for reference purposes. For ease and to cut down on repetitions, "standard" will mean the NFPA 1983 standard.

ENGINEERING ROPE

Two primary considerations when rope is engineered are what material to make it from and what method of construction to use. The word *fiber* will refer to the material component of a rope, whatever its specific characteristics. The first category of materials available to man were the natural fibers from plants and trees. For centuries, humans have woven ropes from a long list of regionally available materials. Today, natural fibers aren't even allowed in the fire service or in any other life safety situations because of their limitations. Natural fibers such as cotton, sisal, jute, hemp, and manila were some of the common materials once used to lay rope. Manila became popular in the fire service because of its strength and reliability. Found in the abaca plant, it is graded according to the quality of its fibers. That which is derived from the inner portion of the plant has a Grade 1 designation. These plant fibers are limited to about 12 to 15 feet in length by nature; when they are processed, they get cut down even shorter than that. To manufacture cordage, fibers are turned into yarns, yarns into strands, and strands into rope.

The standard requires that life safety rope be of block creel construction. This means that the fibers must run continuously throughout the rope. As you can see, short natural fibers can't be used because one finished rope is actually many little ones joined together. The issue of continuous reliability comes into play. Other references in the standard also knock natural fibers out of the picture. In assessing materials for construction, natural fibers have many drawbacks in terms of life safety use:

1. A rope of natural fibers absorbs everything it contacts. It's one long sponge. Years ago, new rope was soaked in water to soften it up. This action also decreased its breaking strength, which was poor to begin with.
2. Natural fibers rapidly degrade, even when properly stored.
3. A natural-fiber rope attracts mildew because it never really dries out. It can also rot quickly.
4. Natural fibers burn and lose their load-carrying ability more quickly than synthetics.
5. In prior years, the rope in service often wasn't Grade 1, so the incidence of failure was higher.
6. Natural fibers have low resistance to abrasive surfaces.
7. Natural-fiber ropes have low resistance to shock loading.
8. A rope of natural fiber can absorb 50 percent or more of its own weight in water—a real drawback in any rescue.
9. Comparable synthetic ropes are dramatically stronger.

Currently, hemp products are making a comeback in the United States in the form of clothing and other products. Still, natural fibers have disappeared from the mainstream fire service and shouldn't even be used for utility work. Those departments that do downgrade their life safety ropes should identify them appropriately to lessen the chance of an error being made.

Synthetic fibers came out of the great World War II industrial process. Like so many great inventions, they are children of the military. The armed forces were looking for a lightweight yet rugged fabric that would have many applications. In 1938, E.I. du Pont de Nemours & Co. invented nylon, and the rest, as they say, is history. Today, several synthetic fibers are in use in the rope rescue field. Let's look at synthetics in general and their attributes with respect to safety:

1. Synthetic fibers can run continously throughout the length of the rope for greater strength.
2. They don't degrade or age as fast as natural fibers.
3. They allow different engineering methods to be used, resulting in a better finished product.
4. They have greater resistance to rotting.
5. They don't absorb water as readily as natural fibers. This is due to the particular polymer being used, the treatment of initial filaments, and/or a postmanufacturing dry treatment.
6. They melt or burn at higher temperatures than natural fibers.
7. Their strength factor compared with natural-fiber ropes of similar diameter is dramatically greater.
8. They have higher resistance to shock loads.

Some well-known drawbacks of synthetic-fiber ropes are:

1. Loading them over a sharp edge can cause them to fail.
2. Prolonged exposure to sunlight and its ultraviolet rays can damage them or decrease their life expectancy.
3. Contact with acids or alkalies can degrade them.

The synthetic fibers of modern rope construction are grouped into families. All of these families have specific characteristics setting them apart from the others—characteristics that decide the design and manufacture of a given rope.

THE SYNTHETIC FIBERS

Nylon

Let's look at the individual fibers used in life safety applications. The polyamides are the nylon family. This multiuse fiber is used in thousands of products today, ranging from rescue ropes to hunting socks. Nylon is *the* predominant fiber for rope rescues and webbing. This material has many more assets than drawbacks. There are many manufacturers of nylon fibers today since its invention by DuPont. When promoting rescue rope, marketers will list the specifications as nylon or as a manufacturer's chemical formula for its specific fiber. In Europe, you may see polyamides listed as Perlon in Germany and as Grilon in Switzerland. In the early days, Perlon was used quite a bit. Perlon is the name for Type 6 nylon, used in many world-famous climbing ropes.

The attributes of nylon are:

1. It is the predominant material for life safety ropes due to its superior characteristics and cost-effectiveness.
2. It is more elastic than its closest cousin, polyester.

The drawbacks of nylon are:

1. It loses approximately 15 percent of its strength when it is wet, though it will regain that strength when dried.
2. It can be damaged by various chemicals, including acids.
3. It has a melting point of approximately 480°F, a concern whenever high temperatures or direct flame impingement is possible.

When searching through manufacturers' or distributors' catalogs, many times you'll see references to nylon 6 or nylon 6.6. These numbers refer to the chemical formula of the polymer itself. For example, DuPont manufactures nylon 6.6, then classifies the filaments according to type. This further classification denotes the various properties for specific customers, such as manufacturers of rescue rope. To complete this illustration, sometimes you'll see references to such fibers as nylon 6.6 or DuPont 707. Out of the 13 types listed in the DuPont Industrial Filament Yarns tech bulletin, two are often referenced in rope catalogs: Type 707 and Type 728.

Polyester

Polyester is the second most common fiber used in the construction of rescue ropes. Polyester fibers are also known as Dacron™ by DuPont.

Dacron ropes were popular in the three-strand laid version during the late '70s and '80s. This was when pompier belts dominated; before static kernmantle and technical rope equipment became more commonplace in the mainstream fire service. (Kernmantle rope, as will be discussed below, is comprised of a core of many fibers and an outer sheath for form and protection.) At that time, Dacron ropes were purchased from commercial cordage companies rather than traditional rescue rope suppliers. Polyester is a fantastic fabric with many of the assets of nylon but also a few differences worth mentioning.

The attributes of polyesters are:

1. Polyester doesn't absorb water to any degree, being only about two percent hydrophilic.
2. Polyester has a much higher resistance to ultraviolet light and potential sunlight damage.
3. Polyester has better resistance to abrasive surfaces.
4. Polyester has better resistance to acids.

The drawbacks of polyester are:

1. Polyester's strength is approximately 10 percent less than that of nylon.
2. Polyester has approximately half the shock-absorbing capacity of nylon.
3. Polyester can be damaged and weakened by contacting alkali chemicals.

Several rope manufacturers produce rescue ropes made with nylon kerns and polyester mantles, drawing from the best attributes of both varieties.

Vectran HS® is a high-performance filament yarn spun from Hoechst Celanese's Vectra® liquid crystal polymer. This is the newest of the fibers available, and it's beginning to appear in the rope rescue field.

The attributes of Vectran are:

1. High strength.
2. Low moisture absorption.
3. It doesn't creep.
4. It retains its properties over a broad temperature range.
5. It has extraordinary chemical resistance.
6. It is easily knotted.

The drawbacks of Vectran are:

1. It costs approximately *15 times* as much as raw polyester fibers. Presently this fiber is being used in the production of accessory cords.

The Polyolefins Family

Polypropylene and polyethelene are the lightest of the synthetics discussed so far. They are divided into either monofilament or multifilament fibers. Modern fishing line is constructed from multifilament fibers, as are these poly ropes.

The attributes of the polyolefins are:

1. They don't absorb water.
2. They float on water.
3. They resist degradation by many chemicals.

The drawbacks of the polyolefins are:

1. They have low tensile strength.
2. They rapidly deteriorate from exposure to sunlight.
3. They knot poorly.
4. They have only about 60 percent the energy absorption of nylon.
5. They have low melting points—approximately 200°F.
6. They have poor resistance to abrasive surfaces.

The bottom line is that this material shouldn't be used for life-dependent activities, including any technical rescue operation where the rope will be supporting a life. This material is popular for water rescue situations where rescuers are supported by a primary flotation device rather than the rope. If in doubt, use a rescue-grade rope.

The Aramid Family

Kevlar™ by DuPont is the latest material to enter the rescue service field. Most people associate this material with bulletproof vests. Southern Mills of Union City, Georgia, also blends it with PBI™ to make turnout gear.

The attributes of Kevlar™ are:

1. It has a very high strength-to-weight ratio.
2. It has good stretch resistance.

3. It doesn't creep.
4. It has excellent tension-tension fatigue life.
5. It is usable over a wide temperature range.
6. It is electrically nonconductive.
7. It has outstanding environmental and chemical resistance.
8. Pound for pound, it is five times as strong as steel, having one-fifth the weight of steel in air and 1/20th the weight of steel in water.

The drawbacks of Kevlar™ are:

1. It has poor resistance to abrasive surfaces.
2. It has poor shock-absorbing qualities.
3. Bending it to small radii, as in tying and untying it, easily damages the fibers.

Presently, Kevlar™ is used in small quantities for the construction of accessory cords for chocks. Usually it's combined with another material such as polyamide (nylon) in the finished product. It's still a newcomer in terms of acceptance and long-term performance in the rope rescue field.

ROPE CONSTRUCTION

The actual method of construction is the second consideration in the manufacture of rope. While the material is critical, so is the intended use of the finished product. This is where you, the customer, must buy rope based on its attributes if it is to safely perform a given task.

Laid Construction

Laid construction was the predominant method for many years, using both natural and synthetic fibers. When the fire service purchased hemp or manila rope, it was a three-strand right-hand laid rope. This method is still used and popular for commercial cordage. It is also known as twisted, plain, or hawser-laid rope. I'm told that the current term *hawser* is an indication of extremely large-diameter rope, such as is used in the marine industry. Anyway, I grew up knowing it as hawser laid—like many terms out there, it has taken on a life of its own. Modern synthetic twisted ropes are available in a soft or hard lay. The soft lay is indicative of commercial or nonrescue-grade ropes, being supple and easily knotted. The hard or mountain lay is a tightly

Laid Rope

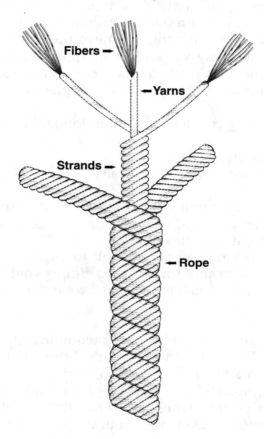

twisted rope, providing some resistance to abrasion. It isn't easy to tie knots with it, and they need to be backed up. The brand manufactured by Plymouth, called Goldline™, was quite popular well into the early '80s. U.S. military personnel use a similar rope called greenline for tactical air assault operations. This product is still available in military supply catalogs.

Although sport climbing has had a tremendous influence on the rescue field, so too have the cavers. Many belong to the National Speleological Society, and explorers such as these have developed special techniques while pushing for technological advances. Among their needs was a rope that could support them on long vertical descents. In caving, such descents can reach more than 1,000 feet. To meet such distances, single-rope techniques had to be developed and refined. Single rope means just that—no belay or backup. The mountain lay was decently suited for the abrasive environments that cavers encounter; however, its drawbacks were accentuated by the long drops. Such rope passing through a descending device causes spinning and kinking and makes for a tough time at the bottom of a rappel.

Although it took many years for it to appear on the scene, cavers sought a replacement for the laid rope. Let's examine some of the handling characteristics of laid rope when it is made out of nylon rather than natural fiber.

Common handling characteristics of nylon laid rope are:

1. It stretches under load.
2. It untwists when loaded and causes a spinning or kinking reaction.
3. When its load-carrying fibers are exposed to the environment, abrasive surfaces, and heat-generating mechanical devices, the rope's integrity can be compromised.
4. Nylon hard-laid ropes are difficult to manage for transport and deployment because their stiffness limits your options.
5. More friction is created by laid ropes due to their bumpy exterior.

In the final analysis, laid ropes are predominantly used for utility work rather than life safety purposes. One of the problems experienced by personnel who still use ladder belts or the pompier-style friction device is that right-hand laid rope untwists itself. There have been occasions when this untwisting has opened the gate of the descending device, disconnecting the member from the rope. This has been remedied by using left-hand rope and aggressive training. By and large, kernmantle ropes are what are predominantly used today. The largest user of laid ropes for fire service rescue operations is the Fire Department of New York, which uses a 9/16 left-hand laid nylon rope with a snap link at each end, packed into a rectangular carrying bag. FDNY employs specific techniques and hardware for lowering members down to victims.

Plaited Construction

With plaited construction, the specific terminology involved seems to depend on either the manufacturer or what region of the country you come from. Essentially it is constructed by weaving eight strands together to form a rope.

Plaited construction's common characteristics are:

1. It forms a very supple rope.

2. Its load-carrying fibers are totally exposed and subject to damage.
3. It stands a higher chance of getting snagged by abrasive surfaces, a condition known as "picking."

Plaited Rope

Because of its drawbacks, plaited rope isn't suitable for life safety applications.

Braid on Braid

This rope became the replacement for the mountain lay used by spelunkers. It was adopted for use in the 1960s because of its high strength and low-spin characteristics. Remember that the spin of the laid rope used to be an operational nightmare. In those days, braid on braid was called Samson rope, after its manufacturer.

Double Braid

This rope is essentially one braided rope over another. It is considered an early version of the kernmantle rope. It became popular with the fire service along with the nylon- or Dacron-laid rope in the late '70s and early '80s. At that time, it wasn't uncommon to see 3/4- to 1-inch-diameter ropes with eyehooks spliced into the ends. They still appear on apparatus in my region of the country.

Double-Braid Rope Showing the Sheath and Braided Core

The common handling characterisitics of double-braided rope are:

1. It's supple.
2. It's prone to picking.
3. It's prone to core and sheath slippage.
4. Contaminants are not easily cleaned from it.
5. Its load-bearing fibers are exposed to damage.

Although suitable for life-saving operations, double-braided rope has largely been replaced by kernmantle varieties.

Kernmantle Rope

The term *kernmantle* is German in origin, kern meaning core and mantle meaning sheath. Both words denote the two parts of the rope's construction. This is currently the predominant type used in caving, sport climbing, and rope rescue operations in a variety of disciplines, including the fire service. Intrinsic to any discussion of kernmantle rope are the terms *static* and *dynamic*. At issue is the elongation ability of any given rope. Both types are important to rope rescue operations and offer different characteristics to suit different needs. Static kernmantle was originally used for caving in the United States. Dynamic kernmantle traces its origins back to sport applications in Europe. Regardless of variety, kernmantle ropes manufactured in accordance with particular standards provide a high degree of safety to the user. Many times, people will arbitrarily refer to a rope as being either dynamic or static, causing a controversy over terminology. Since the predominant rescue rope is "kernmantle," and that's all you use, then the lingo is probably okay. All you need to remember is that all of these ropes have elongation characteristics, some more than others.

Static Kernmantle: Like any rope, static kernmantle was created with a specific purpose in mind. As mentioned above, spelunkers were looking for a better rope to accomplish their single-rope techniques during long descents. They needed a rope that had high resistance to abrasion, low stretch, and low spin. Prior to its use in the rescue field, this rope was even known as caving kernmantle. It was pretty much exclusive to the sport from the late '60s to the early '80s. Then, in 1980, two FDNY firefighters fell to their death when their lifeline broke during a fireground rescue. This event and a revised standard, published in 1985, brought fire service rope problems to the fore and evoked a positive change. Of all ropes, static kernmantle is now considered the best-suited for mechanical advantage systems and most fire service rope operations. Due to the influence of European dynamic rope, static kernmantle diameters and lengths are expressed in metric units. Ropes are sold as 9 mm or 11 mm diameter and 45 or 50 meters in length. After much debate and consultation among manufacturers, instructors, allied professionals, and union interests, a consensus has specified life safety rope to have a breaking strength of 9,000 lbs. and a working-to-breaking safety factor of 15. At the time, kernmantle rope wasn't mainstream, however, and the breaking strength requirement was met by using 9/16- to one-inch laid nylon rope—the commonly used 11 mm (approx. 7/16 inch) static kernmantle rope wasn't strong enough. I can remember two problems arising when trying to purchase 9,000-lb. static kernmantle at that time. First, much of the mountaineering hardware used by the fire service was designed for ropes of 7/16 inch and smaller. Second, the only kernmantle rope available in that strength range was the 16 mm (approx. 5/8 inch), which had a rating in excess of 13,000 lbs. This rope went well and above the requirement and filled our needs. Operationally, this 5/8-inch rope was heavy and a chore to tie in knots. Also, it was difficult to find suitable and inexpensive hardware for it. The standard, once it came out, listed the requirements for a one- and two-person load only, not a specific diameter (although it does give minimum and maximum circumferences). Eventually manufacturers reengineered the 1/2-inch rope to meet the 9,000-lb. requirement. I remember calling a manufacturer and asking him what his company had done differently. The spokesman's response was that they'd "stuffed more fibers into the sheath."

As an aside, the metric measurement is currently being replaced by inches due to the large demand for static rope and almost exclusive production by U.S. companies.

Dynamic Kernmantle: This type has been around longer than static kernmantle. Europe's infatuation with its beautiful mountain ranges required a special rope to provide sport climbers a higher degree of safety. Besides its greater ability to stretch, it has a lower resistance to abrasion than static kernmantle. Because it has been engineered to absorb energy by stretching, it is used in situations where shock loading might occur—obviously this is better than having the human body bear the full force of a fall. Dynamic rope has limited applications on the fireground at this time, either because of prejudice, lack of knowledge, or needs that are variant to its characteristics. Many of the evolutions performed on the fireground require a low-stretch rope for a variety of reasons, and static kernmantle remains the reigning fire and rescue rope. Generally speaking, dynamic rope would be used in situations where a firefighter might fall or need to be belayed.

Yarn filament purchased by the beam.

Construction of Kernmantle

There are several U.S. companies that now manufacture dynamic as well as static kernmantle. The overall construction process is about the same. At the beginning is a filament yarn that comes by the beam, a large roll containing the smallest unit of fiber to be used. The core section is constructed into bundles from the base yarn, and this is where each particular style gets its operating characteristics. Static kernmantle's elongation runs at about 1.5 percent with a 200-lb. load to just under 20 percent at yield. Dynamic kernmantle's primary trait is that of high stretch or energy absorption, important to any belaying evolution. Elongation of this type runs at about six percent at a 176-lb. load to just under 55 percent at yield, depending on the manufacturer. The core accounts for about 70 percent of the rope's strength and the sheath for the other 30 percent. These ratios can fluctuate depending on the manufacturer and how a given rope was engineered. In many instances, the fibers used in construction are pretreated at the point of manufacture for added water repellency prior to being braided. Special lubricants are sometimes used to reduce friction. In general terms, static cores are constructed in parallel fiber bundles, although new advances in technology are allowing static ropes to be manufactured in ways similar to dynamic types. Both dynamic and static cores are braided or twisted in a specific pattern of Z's or left-handed twists and S's or right-handed twists. This creates a balanced and torque-free end product. The

Dyed yarn on bobbins, ready to be processed.

core fibers are usually white, since this helps control cost. Also, they can easily be identified if they pop out of the sheath. The mantle can come in a whole host of colors from plain white to motley. Static ropes typically are of one solid color and a contrasting tracer. Dynamic ropes tend to come in three to four colors. This is one way departments can identify their ropes for members who don't quite understand the difference.

During manufacture, a basic white filament yarn can be colorized in one of two ways. The first is to add a colorant to it during the original extrusion process; the second is to dye it afterward. Dyeing is the more common practice. The yarn is sent out to be dyed at a facility that specializes in this process. Obviously, this adds to the cost of the finished product.

Dynamic kernmantle in the making.

The difference in price for a static rope is approximately ten cents a foot.

Next, the fibers are wound onto braider bobbins, small spools that hold the smaller yarn fibers so that they can be set into the braiding machine. Other machines make the core fiber bundles and prepare them for final setup. The core bundles are drawn through the braiding machine while the sheath is woven around it. To construct a rope with multiple colors requires the dyed yarns to be placed in a specific order so as to be properly braided.

Besides type, diameter, and color is the consideration of dry-treating a rope. Increased awareness of what moisture can do to a rescue rope has placed more and more emphasis on treating the core and the sheath. Wet ropes have lower tensile strength. They can freeze more quickly and are more likely to experience failure if shock loaded. Also, water in a rope increases the difficulty of using auxiliary system components, besides adding to the load.

PURCHASING ROPE

Purchasing rope is like buying a car from the top-five list: All are pretty good, but all look and feel different. Rope is the same way, and each manufacturer is looking for your business. I don't buy rope that I can't try out first. There are about five or six major manufacturers in the United States who produce static ropes. I always call for a sample before making a purchase. Most have sent me a good 75- to 100-foot section to rappel on. This gives me a feel for the rope's handling characteristics, and these sections of rope tend to be too small to sell, anyway.

To begin, you'll want to compare specifications, such as:

1. Materials used in the core and sheath.
2. The weight per 100 feet of rope. A heavier product may indicate greater strength or higher abrasion resistance.
3. Sheath thickness can affect abrasion resistance. Some ropes are designed to take more abuse; however, the trade-off is a stiffer rope.
4. Some ropes are coated in an effort to increase their abrasion resistance.
5. Do you need or want colored rope? Many are white with a tracer for contrast. As mentioned above, colored ropes cost more per foot because they have to be dyed.
6. What length do you need? Rescue rope can be cut to almost any reasonable length. There's at least one U.S. company that can make a 3,300-foot spliceless rope if you're in the market for it.

Until recently, dynamic ropes were purchased primarily from European manufacturers. Their lengths are generally 45 or 50 meters (approximately 150 and 165 feet, respectively) and are 11 mm (7/16 inch) in diameter. These are considered one-person ropes under the standard. Today, dynamic kernmantle is available from several U.S. manufacturers. These companies produce not only quality and features that rival their European counterparts, they also produce standard lengths from 45 to 200 meters without additional cost.

ACCESSORY CORDS

Accessory cords are small-diameter cords primarily used to form grab devices (prusiks) that work on rescue rope. They are also used for chock slings, equipment tie-downs, tag lines, and many other nonlife safety applications. Cordage under 8 mm is generally considered accessory cord. As with rescue rope, accessory cords come in many sizes, options, and handling characteristics. When you read marketing catalogs, you'll see several different terms used in connection with them. These aren't just little kernmantle ropes; they aren't rated for self-supporting life safety applications. Terms such as accessory cord, prusik cord, and rescue cord designate them in the catalogs. A vast majority are of nylon static kernmantle construction and are available in various diameters and sheath thicknesses.

When used to tie a prusik knot on a rescue rope, these sheath thicknesses influence the grip on the rope. In general, the stiffer the cord, the less effective the grip; the softer or more pliable the cord, the more effective the grip. Consequently, there is a trade-off in the manufacturing process between strength, wear, and suppleness.

WEBBING

Webbing is the last member of the fabric family to discuss. Webbing is also known as tape, flat rope, and sometimes slings. It is a common and highly versatile component in many rope rescue systems and is used to construct individual harnesses, presewn slings and runners, hastily tied harnesses, pretied slings, and much more. Webbing for rope operations comes in two broad styles, flat and tubular. A quick look and you'll find them both to be flat in appearance. It's when you look at a side view or roll them in your fingers that you'll be able to tell that one is solid and the other hollow. Flat webbing resembles seat belt construction, except that it's stiffer and difficult to tie into knots. Tubular webbing is supple and easy to work with; therefore, this is the type most widely used for rescue operations. The construction method for tubular webbing varies and needs to be considered. Only buy it from a reputable dealer. There are three methods used in its construction, the first being an older style of chain or edge-stitched webbing. The second is spiral or shuttle loom webbing. The third is called needle loom. The problem with the older style of chain webbing is its low resistance to abrasion and the likelihood that it will unravel once the stitch is exposed. This type is pretty much extinct among the major suppliers of rope rescue equipment. Spiral-constructed webbing resembles an enclosed tube that is flat in appearance but hollow in the center. However, the spiral is being replaced by the needle loom, since the spiral or shuttle loom machines are no longer available. All new machines produce the needle loom style. The needle loom resembles the chain loom but is actually superior to the shuttle loom type.

Webbing is available from 1/2 inch to three inches in width in both varieties, depending on the manufacturer, in both flat and tubular styles. One-inch tubular is perhaps the most widely used type in the rescue services field, although a lot of two-inch webbing is used. Many anchor systems are built with it because webbing takes the brunt of the abrading surfaces and it's expendable. Webbing is also relatively cheaper than rescue rope and is considered disposable so as to prolong the life of the rope. When purchasing webbing, make sure that it conforms with the applicable mil-spec; for example, a one-inch tubular meets mil-spec type Mil-W-5625 from the manufacturer.

QUALITY CONTROL

Unbeknownst to many, a multitude of influences can affect the quality of all rope rescue equipment. To protect the user, manufacturers that produce this equipment must follow a series of standards, administrative guidelines, and regulations. Let's define those terms:

1. A standard is a document put together through a consensus of individuals sharing interest in a specific field. Standards are created to fill a gap and establish some rules as a basis for operations. In the case of NFPA 1983, the standard is essentially a document that lists performance criteria for components used in the fire service's rope rescues. A standard doesn't have the force of law, since many come from nongovernmental associa-

tions. A standard may be used against you in litigation, however. You may be held to one because that is what your peers use in performing their duties. A standard may have the force of law if it's adopted by your organization or incorporated by state law.

2. Administrative guidelines tend to be methods, procedures, processes, and the like that are promulgated by associations, industry trade groups, and government agencies. They can be considered standards; however, they are usually a small piece of a larger document.

3. A regulation is issued by a government entity and is considered a law, carrying the force of compliance and penalties for noncompliance. An OSHA regulation would be a good example.

Following are some organizations that directly influence the technical outcome of components used by the fire service for rope rescue:

NFPA
National Fire Protection Association
Batterymarch Park
Quincy, MA 02269-9101
(800) 344-3555

NFPA 1983, *Fire Service Life Safety Rope and System Components,* is the document that fire service personnel look to when purchasing this type of equipment. Contained within this and many other NFPA standards are references from other associations, agencies, and trade groups. Such references are technical attributes that ensure quality control or minimum performance.

UIAA
Union of International Alpine Associations
c/o American Alpine Club
710 Tenth Street, Suite 100
Golden, CO 80401
(303) 384-0110
E-mail: 10064.3042.@compuserve.com

This organization establishes performance criteria for equipment used by moutaineers. This encompasses hardware, harnesses, and dynamic ropes.

ANSI
American National Standards Institute Inc.
11 West 42nd Street
New York, NY 10036
(212) 642-4900

This organization approves standards from other organizations with regard to technical accuracy. Organizations such as the NFPA and the ASTM are accredited standards developers who submit their documents to ANSI for approval as an American National Standard. ANSI is the coordinating body for national standards and is the U.S. representative of ISO. ANSI also sets standards of its own for a variety of occupations and products. ANSI has a strong interest in standards for safety belts, harnesses, lanyards, life lines, and drop lines for construction and industrial use. This directly relates to equipment used for confined space rescue.

ASTM
American Society of Testing and Materials
100 Barr Harbor Drive
West Conshohocken, PA 19428-2959
(610) 832-9728

This organization is involved with setting standards for materials and techniques used by a wide variety of industries. It establishes a standard by using a consensus of all affected parties. This includes designers, manufacturers, marketers, and customers. Presently the ASTM F-32 Committee on Search and Rescue is working on a rope and hardware standard for equipment used in search and rescue operations.

OSHA
U.S. Occupational Safety and Health Administration
Department of Labor
200 Constitution Avenue
Washington, DC 20210
(202) 219-8191

This organization is involved in workplace safety. OSHA doesn't have any regulations specific to fire service rope and ancillary equipment presently; however, some of its regulations may preclude using regular rope techniques in certain situations. 29 CFR 1910.146, the confined space regulation, is one that comes to mind. Check with your local federal or state OSHA or equivalent for interpretations.

TESTING

To ensure the safety of the user, many of the primary components critical to life safety are tested against the applicable standard. For example, NFPA 1983 lists the minimum breaking strength for one- and two-person ropes, which are 4,500 lbs. and 9,000 lbs., respectively. You must read the fine print when purchasing rope, since some manufacturers list the average and not the minimum.

The Three-Sigma Process, found in the 1995 revision of NFPA 1983, spells out how to determine minimum breaking strength. It is determined by subtracting three standard deviations from the mean of five test samples. Manufacturers will also test their ropes over an abrading surface to see how many repetitions it will take to break through the sheath. This is a good marketing tool.

The UIAA standard tests dynamic ropes for overall strength and the number of falls it can handle. Anytime fall factors approach 0.25, you need to work with dynamic rope. The test assesses how many falls a rope can take under simulated conditions. In this test, a sample of dynamic rope is attached to an 80-kg (approx. 176 lbs.) weight to simulate a climber and is then dropped. The rope pulls tight over a 30-degree edge, which simulates a carabiner located below the climber. To receive the UIAA certification, the rope must withstand at least five drops without breaking. You'll see manufacturers' catalogs listing ropes that withstood nine to twelve falls. This doesn't mean that you can arbitrarily fall that many times and not have a failure. Depending on conditions, any rope can fail on the first occasion of dynamic loading.

ROPE MANAGEMENT

Rope management encompasses readiness, deployment, and on-scene handling of rescue ropes. All ropes, webbing, harnesses, hardware, and other items should be bagged or stored in a way that provides quick access and easy deployment. Time-honored methods are to store ropes in bags and to clip carabiners to an equipment sling. The items should be stored in a clean, cool, and dry area of the station, and the area should be of limited access so that the items don't disappear.

When stored on fire apparatus, the equipment's compartment should likewise be clean, dry, and oil-free. It should hold only the rescue package. Don't store the package in the same compartment as the hydraulic spreaders or vent saw. The cleaner the equipment stays, the longer and more reliable a life it will have. Readying rope for deployment is pretty

(Top) Harnesses stored and ready to go. (Left) Gearbags and an SAR system. (Right) Long lengths of rope stored on spools for quick deployment.

simple these days, since kernmantle is easily bagged. Over the years, many different methods were used to get rope from one point to the next. Today, the primary method of storage and deployment is simply to stuff the rope into a rope bag. Stuffing the rope on top of itself, layer by layer, works quite well, and usually it will pay out cleanly. It's recommended that the first end in the bag have a figure 8 stopper knot tied into it, since this will hopefully prevent the rope from slipping through a device once the end is paid out. Only one rope per bag is allowed, since this keeps it ready and easily transported. Once you get to where it is needed, tie off the working end to a substantial object, then throw the rope bag down to the ground. I've seen rope bags thrown over the edge without being tied off—what's the problem there?! You should communicate to anyone below where the bag is going to be thrown. You want to keep the rope's path away from the victims, since they may jump for it or otherwise do something to stall the rescue operation; also, you don't want to hurt them. You can yell, transmit on your portable, or do

Tie a figure 8 stopper knot in the end of the rope that will be placed in the bag first. The other end should have a loosely tied figure 8 on a bight plus a safety. Some teams leave the end untied and tie the knots on-scene.

When stuffing rope, have someone hold the bag open to make the job easier.

both to broadcast your intentions. Yelling "rope" and then waiting a few seconds before making a drop is an accepted practice, and if someone yells back "hold" or "no rope," then you have time to stop yourself. Of course, dropping a rope and bag is hard as hell on the equipment, anyway. Not dropping it under most circumstances eliminates the need for most of these precautions and makes for a safer work environment. Lower what you need and keep the rest on top for use.

Managing different ropes during certain evolutions can be difficult and presents safety concerns. Knowing the difference between static and dynamic kernmantle and how each fits into an operation is a good start. When multiple lines of rope of the same construction are used in an evolution, you have to differentiate them in some way. This is where different colors are valuable.

ROPE CARE

Proper care of ropes and other fabrics is critical for successful rescues. You need all of those life-dependent components ready and defect-free. I prefer the term care rather than maintenance because care connotes a more nurturing, all-encompassing attitude. Care starts from the day you purchase the product, put it in service, use it, and prepare it for future use. Your life actually depends on how you treat that or any other component. Here are some general points to remember:

1. Never step on a rope. This can force sand and other contaminants into the core and break down the fibers.
2. Protect rope from prolonged exposure to sunlight.
3. Protect rope from sharp or abrasive surfaces.
4. Protect rope from heat, flames, and chemicals.
5. Protect rope from shock loading.
6. Protect rope from contacting other fabrics. Friction between ropes or any other fabric can cause it to melt, burn, and fail.
7. All ropes, webbing, and accessory cords must be washed and dried according to the manufacturer's guidelines.

Washing Rope

Many of the modern fabrics used to make rope can be washed in a solution of mild soap and cool water. Rinse the rope thoroughly several times to remove the soap. Don't be afraid to wash the rope—minimal strength is lost while it is wet, and it regains all of it when it is dry. Never use bleach or stiff scrub brushes to clean rope or other life safety fabrics.

Drying Rope

Don't hang the rope, webbing, or accessory cords in direct sunlight, which can deteriorate the fabric's strength. Don't use mechanical heating devices. Air drying is best and can be done in most fire stations.

Frozen Rope

If ropes become frozen on the scene, don't force them back into their bags. Instead, lay them, secured in long lengths, in the hosebed or along the top of the ladder truck for transport back to the station. Thaw them out in a grease- and oil-free area of the apparatus bay. Forcing frozen ropes into their storage positions before they are thawed will damage them.

INSPECTING ROPE

Slowly run rope, webbing, and accessory cord through your hands to inspect it. Feel the fabric inch by inch from one end to the other. Be alert for any discoloration, depressions, core slippage, abrasions, and flat spots. The inner fibers of kernmantle, because of the way it is constructed, aren't visible as they would be in twisted or laid ropes; therefore, you must carefully inspect the outer sheath as a skin. Feel it for flat spots, depressions, core slippage, and lumps on the inside. Any or all of these indicators generally mean that there is some internal fiber breakage. It is commonly held that if the sheath is intact and damage free, so is the rope's core.

There are occasions when some of the aforementioned indicators need further analysis. One way to confirm fiber breakage and retraction is to stretch out the rope and tension it with the equivalent of a one-person load. If a rope under tension displays a decreased diameter or localized reduction, then take it out of service. Also, cut it to expose the damage. No doubt you'll feel better about your decision once you've been able to see the defect firsthand.

Some common visual indicators that need to be addressed immediately are:

1. Sheath worn through to the core.
2. Core fibers popping out of the sheath, either with or without sheath damage. Fibers popping out occurred in the past when cores were spliced rather than of block creel construction. Contact the manufacturer, especially if the rope is fairly new. It may replace the rope free of charge for reasons of quality control and company reputation.

3. Any discoloration.
4. Shiny markings that indicate heat or friction burns. If you have these, then most likely you have flat spots—the normally round rope is flat and melted.

The issue of sheath fuzziness comes up often. This fuzz is created when the individual fibers of the sheath are broken during normal use. This relates directly to the abrasion resistance of the sheath and the rope's overall life expectancy. Ropes that are used for rapelling or that operate under abrasive conditions tend to build up this fuzz, unlike ropes that are used in pulley or belay situations. Generally speaking, a little bit of fuzz spread out isn't an automatic ticket to the retirement home. Presently, there is no guaranteed method to determine when enough is enough, though retiring a rope after a one-time lifesaving rescue is a good idea.

All rope, webbing, and accessory cords must be clean, dry, and inspected prior to being placed in ready status. Some administrators and chief officers feel that the ropes purchased should last forever. They don't have to swing from them, so it's not a personal issue that affects them. Good-quality rope is expensive, but so is human life. Many fire departments downgrade used ropes to applications such as knot-tying training or utility use. All cordage must be used and cared for according to the manufacturer's guidelines.

RECORD KEEPING

Record keeping is an important administrative task that must be done for accountability, legal, and budgetary reasons. Keep current inventory records and complete, accurate history cards on every piece of equipment. This includes each length of life safety rope, which must be tracked from the day it is purchased to the day it is destroyed. A history card can be put in a resealable clear plastic bag and kept in the rope bag. Once the card is full, it should be turned over to the individual in charge of the equipment tracking, who should store it and issue another card.

Individual lists of what is carried on each apparatus should be available so that personnel can cross-check the status of equipment as a backup. Such records can be kept on cards, on specially designed forms, or in a computer.

Each department must develop a system suited to its needs. The cards, along with an inspection, can help determine the serviceability of equipment. When in doubt about the integrity of any item, exclude it from life

```
┌─────────────────────────────────────────────────────────────────┐
│                    HISTORY OF RESCUE ROPE                         │
│                                                                   │
│  Assigned apparatus :_____      │
│  Date of Purchase: _____/____/_____    Purchased from: _____ │
│  Rope Size:  dia:_____  length: _____  Static_____  Dynamic_____ │
│  ───────────────────────────────────────────────────────────     │
│  Date of use: _____/_____/_____   Officer in charge: _____  │
│  Manner of use: _____  │
│  Condition after use: _____  │
│  ───────────────────────────────────────────────────────────     │
│  Date of use: _____/_____/_____   Officer in charge: _____  │
│  Manner of use: _____  │
│  Condition after use: _____  │
└─────────────────────────────────────────────────────────────────┘
```

safety operations. A tracking system and inspection program will also be extremely important should liability issues arise.

EDGE PROTECTION

Edge protection refers to techniques that prevent rope from contacting abrasive or sharp surfaces. Ropes and other cordage need protection because destructive surfaces can weaken or even cause immediate failure. Rope requires protection whether it is in a static situation (not moving) or a dynamic one. An example of a static situation would be rope used for a rappel, whereas a dynamic situation would be the line in a pulley system. In either case, the common denominator is that the rope is loaded and bending. One of the most common reasons for rope failure is mechanical damage caused by sharp surfaces. Protection of rope and other cordage is accomplished with a variety of techniques and equipment.

When we talk about abrasive surfaces, we mean hazards such as masonry, cement, rocks, roof shingles, and other ropes. When we talk about sharp surfaces, we mean glass, metal edges, narrow or pointed rocks, and the like. Equipment to provide protection against such hazards can either be store-bought or fashioned from common items as needed. Store-bought equipment includes commercial rope pads, rope guards, metal edge rollers, pulleys, A-frames, and tripods. Improvised equipment includes canvas rope pads, fire hose rope pads, metal fire hose rollers, gear or carry bags, and turnout clothing. Even carpeting, bedding, and draperies can be used in an emergency for edge protection.

The operating environment dictates the type and level of effort required. Individual situations will be covered in the applicable chapter as specific rope techniques are discussed.

STUDY QUESTIONS

1. Which NFPA standard covers the rope and system components that firefighters use during rope rescue operations?

2. What term indicates that fibers run continuously through a rope?

3. Name the two most popular synthetic materials used in constructing fire service ropes.

4. What are some of the common handling characteristics of nylon laid rope?

5. Which rope construction method is best suited for mechanical advantage systems and most fire service rope operations?

Chapter Five

Hardware

The term *hardware* encompasses many of those connecting, grabbing, and friction devices critical to an operation. Whereas rope and webbing are considered software, carabiners and descenders are considered hardware. A wide variety of both are needed for any system. This chapter will focus on many of the devices available and their use.

CARABINERS

Carabiners are various-size connecting links that are heavily used in many systems. Generally, they connect fabrics together, such as rope

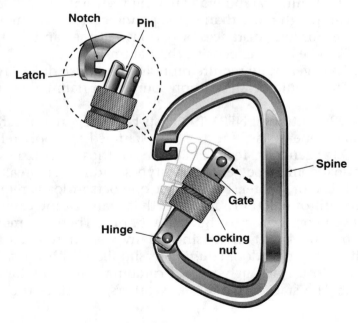

to rope or rope to webbing, to prevent heat damage. They can also connect other hardware to a rope. Carabiners are constructed of either aluminum or one of several types of steel. Like anything else, each material has its attributes and drawbacks.

The attributes of aluminum carabiners are:

1. They're lighter weight.
2. They require minimal maintenance due to rust resistance.
3. They're cheaper than steel.

The drawbacks of aluminum carabiners are:

1. They're generally not as strong as steel models of the same size.
2. Heavy shock loads may permanently distort or completely break an aluminum carabiner.

The attributes of steel carabiners are:

1. They're more rugged and stronger than aluminum.
2. They may tolerate shock loads better than aluminum.

The drawbacks of steel carabiners are:

1. They tend to be much heavier than aluminum. Lightweight aluminum carabiners can weigh several ounces; steel models can weigh more than half a pound each. Carrying large quantities for long distances or periods of time can be fatiguing.
2. They're more expensive than aluminum.
3. They generally require more maintenance due to rust.
4. They could be an ignition source in a flammable atmosphere.

The 1995 edition of NFPA 1983 lists two categories of carabiners, personal and general use, and they are stamped appropriately (P for personal, G for general). Team carabiners, or those that may support two-person loads, must be general-use types and have a breaking strength of 9,000 lbs. or greater. Personal or one-person loads require a breaking strength of 6,000 lbs. It is true that many of the carabiners in use don't even come close to the 9,000-lb. rule. There are many 5,500- to 6,000-lb. models out there, since many were purchased from mountain climbing or outdoor equipment suppliers. This doesn't mean that you have to throw out your old carabiners and purchase new ones, however. The standard, like many others, describes the performance

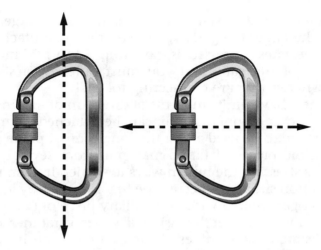

A carabiner is strongest along its major axis.

A carabiner is weakest along its minor axis.

criteria for all new equipment to be purchased from the effective date of the document. Just make sure your future purchases comply with that section of the standard. Whenever any standard's criteria are increased or upgraded, there tends to be some real-world concern prompting the change. In the meantime, there's no reason you can't provide a higher degree of safety with your current inventory. Doubling up your locking carabiners instead of using a single one for two-person or shock-loading situations is a good idea.

The next consideration is the shape of the carabiner, as well as the gate. Carabiners come in a variety of shapes and are tailored to different applications and manufacturers' selling points. The first ones, used in mountaineering, were oval in shape and nonlocking. As time went on, the snap link evolved, taking advantage of modern technological design. Carabiners are intended to be used along the longitudinal axis, since this provides the highest degree of strength and function. Avoid lateral or minor axis loading, since this can cause a failure.

Some common shapes of carabiners are the standard, oval, D, gym D, offset, and pear. Different manufacturers have gone a step further, combining particular shapes and offering them in several sizes. In the past, some weren't suitable for certain operations because of their shape and how stress would affect them. Also, the actual opening of the gate wouldn't allow them to get around obstacles such as stokes stretcher rails. Today, you can get almost any shape you want in a variety of sizes to suit the task at hand.

Carabiners come in standard, large, and extra large, and the larger the size, the larger the gate. When you read manufacturers' literature, you will see gates expressed as, say, from .74 to 1.25 inches. This is the actual gap of the gate, and you must purchase a size that can get around whatever you're connecting to.

Carabiners have limits in terms of direction of loading and the number of fabric components that can be attached. Remembering that carabiners are to be used along the longitudinal plane, there are times that multiple pieces of fabric may pull from several different directions, thus stressing the link in ways for which it wasn't designed. Two such common situations occur. The first is three-way loading from harnesses that get secured with an auxiliary piece of fabric. This situation may be handled either by using a wider carabiner or a screwlink. Screwlinks are close relatives of the carabiner and have limited applications at this time.

The other situation that comes to mind is when multiple anchor points come together. If you connect all the attachments to a common carabiner, distortion or failure of the link may result. This hazard can be eliminated by using an anchor plate or by narrowing the angle of the converging anchor points.

Your inventory of carabiners should include a variety of different shapes and sizes in both steel and aluminum.

DESCENDERS

The Figure 8 Family

These are devices that provide friction to lower personnel and equipment. They can be used in a fixed or moving brake configuration. Depending on local terminology or practices, the devices that first come to mind are those in the figure 8 family. Such devices are nicknamed figure 8, rescue 8, rappel ring, and 8 plate. They can be constructed of either aluminum or steel and come in a variety of sizes and shapes. The basic figure 8 has evolved over the years from a lightweight sport-climbing model to a heavy-duty device with ears. It is used in rappelling and lowering operations. There are two schools of thought regarding earred versus nonearred descenders. Some prefer the nonearred version if they are setting up a rappel in which they are rolling over an edge. Without the ears on the descender, some feel there is less chance of snagging the edge. Others prefer ears because they can prevent the rope from rolling over itself, creating a rope lockup situation. I prefer the figure 8 with ears for this reason. I've seen situations where

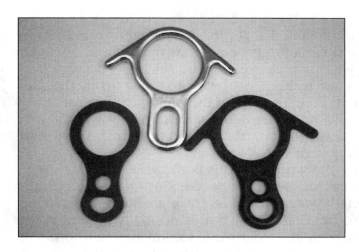

A sample of the figure 8 descenders available.

firefighters have rolled off a flat roof in a tie-low evolution and have gotten girth-hitched with nonearred devices or have hooked themselves to gutters with earred ones. The bottom line is that you must provide enough slack in your rope to clear the edge at all times. Once you master this technique, using either device will be smooth.

Brake Bar Rack

The next descending device, the brake bar rack, is sometimes listed in a category by itself. It has gained much popularity over the past few decades due to its management and control of friction. Previously, the figure 8 family performed the bulk of the lowering operations, including rappels and stokes baskets. The very design principle of the figure 8 creates a situation where the rope can twist and create problems at the far running end. The current practice of many in the field is to limit the use of the figure 8 to short distances of a hundred feet or so and to one-person loads, if possible. The brake bar rack is superior in situations where you have long rappels or lowering distances. The operator can increase or decrease the friction, thus controlling the rate of descent more safely.

Control of the friction is accomplished by a combination of reeving a number of brake bars and increasing or decreasing the distance between them. These bars are made of aluminum or steel and are set up according to the user's liking. Steel bars tend to allow the rope to move more quickly, whereas aluminum provides more friction because the rope actually wears down the metal, leaving deposits on the rope. The current contention of preferences among the experts comes down to a matter of past experiences, the situation at hand, and training issues.

Many out there are still using the figure 8 for all lowering operations.

Many feel that it should be the first choice in emergency egress or rescue operations on the fireground. It's easily transported and relatively idiotproof, provided you maintain your skills. The brake bar rack isn't practical to carry in your turnout pocket or to reeve up in the fast-paced fire environment. Perhaps in time the prejudices against it will change due to aggressive and relentless training. As always, the user must understand the pros and cons of each and train with all of these devices to be fully competent.

Author's note: During the writing of this book, a new NFPA-compliant two-person brake bar rack was introduced. This rack resembles the older-style racks in many ways; however, there are some notable differences. The standard training groove that is normally located at the first bar position has been moved to the second bar position. The manufacturer made this change to get the device to pass the 6,000-lb. test. The next change incorporates a triangular metal tab. This tab is permanently attached to the open side and is slid over to the eye side and secured with a carabiner when the device is attached to a harness. Being a two-person-rated device, all six bars must be engaged and reeved by the rope. This new metal tab ensures that whatever bars are engaged for the descent stay engaged. This new arrangement doesn't allow for the customary adding or removing of bars at the descender's will. Operating these new racks requires specific training and practice.

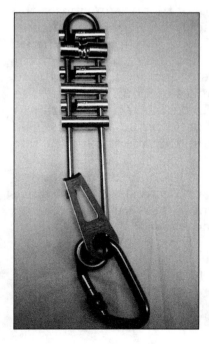

The new-style, NFPA-compliant brake bar rack.

Trying to set up and use the new-style brake bar rack like the older style can be dangerous. Follow the manufacturer's directions for use.

PULLEYS

Pulleys are designed to reduce the friction created by moving ropes. They are most often used to develop mechanical advantage, to change direction, and to reduce or eliminate contact with an abrasive surface.

Rescue-grade pulleys are primarily constructed of aluminum and stainless steel for strength and corrosion resistance.

In the not-too-distant past, achieving mechanical advantage was done by a block-and-tackle arrangement. The block was the predecessor of the modern pulley, and the tackle was the rope properly reeved between the blocks to create the mechanical advantage.

Rope is commonly bent during knot tying and at various points in a system. Bights, round turns, and loops are essential to forming a variety of combinations, depending on the knot. When rope bends in a system, it is usually loaded and turning sharply against an edge, carabiner, pulley, or some other device. Whenever a knot or series of knots are tied in a rope, they weaken that rope a certain percentage. The type of knot is a factor. Anytime a rope bends around an object or system component, in fact, strength is lost. Once a rope is bent around an object, approximately half of the internal fibers are in tension and taking the load, while the other half are in compression and not effec-

Assorted pulleys.

A Block-and-Tackle System

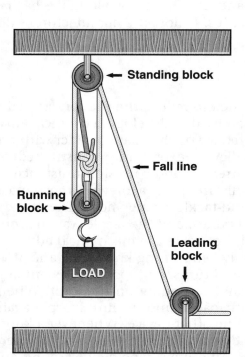

tively taking the load. The exact percentages are highly variable, but drops in line strength and too many sharp bends can cause system failure. Of course, even one sharp edge is too many.

In the past, tensioning a rope or changing its direction was done by running it through a carabiner. It is now known that even a carabiner can diminish line strength and compromise safety. Current techniques include using a pulley instead. A pulley creates less of a bend and provides a movable surface for the rope to transit. Manufacturers have determined that a rope shouldn't bend around any object less than four times its diameter, and this is a minimum ratio. Some manufacturers and instructors suggest ratios as high as 12:1; however, the 4:1 ratio is commonly used. This rule is applicable to many points in any given system.

Rescue-grade pulleys are constructed to handle 1/2-inch and 5/8-inch rope categories. This means that a 1/2-inch pulley is suited for any size rope up to 1/2 inch, and a 5/8-inch pulley for any size rope up to 5/8 inch. You can reeve a 1/2-inch rope through a 5/8-inch pulley, but you can't reeve a 5/8-inch rope through a 1/2-inch pulley. The size of the sheave or wheel is important and relates directly to the 4:1 rule.

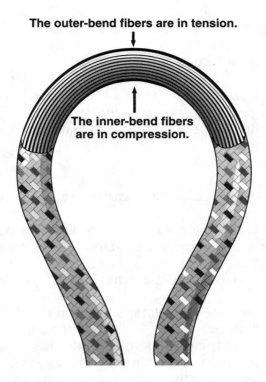

The outer-bend fibers are in tension.

The inner-bend fibers
are in compression.

Manufacturers designate pulleys in terms of the diameter of the sheave. Following the 4:1 rule, you would need at least a two-inch pulley for a 1/2-inch rope. If you were using 5/8-inch rope, a two-inch pulley would be a little too small. You'd need a three-inch pulley at minimum.

Many who use 5/8-inch rope make it a practice to employ four-inch pulleys. They are opting for larger sizes where weight isn't a critical factor, especially when heavy loading is anticipated. This introduces the concept of the tensionless anchor, which is created by wrapping a rope around an object that is at least eight times its diameter. This will be discussed more thoroughly in the chapter on anchoring.

There are two types of bearings germane to pulleys, the Oilite™ bronze bushing and the sealed ball bearing.

The attributes of the Oilite™ bushing are:

1. It has great strength and the ability to handle heavy loads.
2. It can be broken down for maintenance.
3. It's widely available.
4. It's less expensive than the sealed ball bearing.

The principal drawback of nonsealed bushings is that they are more likely to get contaminated by dirt, sand, and grit than the sealed type. The attributes of the sealed ball bearing are:

1. The sealed assembly is permanently lubricated and isolated from the operating environment.
2. It allows the sheave to turn somewhat more quickly than the Oilite™ bushing.
3. It is widely available.

The drawbacks of the sealed ball bearing are:

1. It doesn't handle sudden or heavy loads as well the Oilite™ type. The ball bearings can damage the races when shock loaded, putting a unit out of service.
2. They tend to cost more than the Oilite™-type bushing.

While the sheave diameter relates to the pulley size, the tread size relates to the width of the sheave. The larger the rope and/or the heavier the load, the larger the pulleys you should use.

The sideplates provide the housing for the sheave and make it possible for the pulley assembly to be attached to an anchor or grab device. These plates should be movable to facilitate placing the pulley anywhere along the rope; otherwise, you'll have to reeve it from either end, if that is even possible. When you bring both sideplates together, you align the hole where the carabiner will be clipped, then attach the pulley to an anchor or rope grap. For ease of operations, use pulleys that allow multiple carabiners to be clipped through the top hole. The axles and axle nuts of a pulley should be constructed of steel or stainless steel for strength. The nuts should be rounded and smooth and present a low profile so as not to snag the rope.

Specialized pulleys are available for specific evolutions within a rope system. The knot-passing pulley is designed with a wider tread to handle knotted rope. They are used when two ropes are tied together for long runouts and can be inserted for that particular evolution and removed when completed. When purchasing them, pay attention to the strength ratings, since some are pretty low and may not be compliant.

The prusik-minding pulley is designed to allow a prusik to be attached to a rope without causing binding problems. A normal rescue-grade pulley is designed with rounded edges, which allows the prusik to get pulled into the sheave, stalling the movement. The

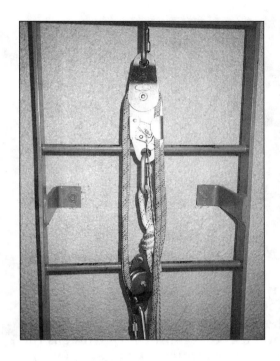

A combination pulley (on top) set up in a 4:1 block system.

prusik-minding pulley has a larger, squared-off bottom to accommo-date the larger bulk of fabric passing through.

The last type of pulley to be covered combines the features of a rescue-grade pulley and a mechanical rope grab. This device is relatively new in the field compared with others yet is already becoming commonplace in confined space rescue. When constructing mechanical advantage sys-tems to pull personnel out of an opening, a 4:1 block system is the typ-ical arrangement. While it does a fine job, you need to add some type of brake or safety mechanism. The combination pulley allows you to per-form the operation without some of the attendant hassles.

ASCENDERS

Ascenders are categorized as either soft or hard. The soft type are prusik knots tied onto a rope to function as rope grabs. The hard type are metal devices that come with or without handles. This section will discuss the two categories of hard-type ascenders. Devices without handles are generally used where two-person loads or other heavy loading is expected. The handled devices are generally for personal ascending, not rescue loads. Therefore, the emphasis will be on the

Assorted low-profile, shell-and-cam mechanical ascenders.

low-profile shell-and-cam device for general or team use.

The purpose of any ascender is to provide a grabbing mechanism that will attach to a rope without causing any lasting effects. The Gibbs ascender is one of the preferred grab devices when constructing a hauling system. The inherent problem that must be overcome is the fine balance in grip—enough grip must be exerted when needed yet be fully releasable so as to take another bite. Dirt, moisture, ice, and other contaminants interfere with the natural bite of the fabric. The Gibbs device consists of an outer shell of either aluminum or stainless steel, plus an inner cam that pinches the rope. Once the shell is slipped around the rope, the cam is placed in position and locked with an aircraft-style quick release pin. Once in place, it won't come off unless the rope fails or the device breaks apart.

The Gibbs ascender is designed to pull from one direction and release from the other. To make things simple, I tell my students to point the arrow toward the load and it will work correctly (see also the section on hauling systems). This series has two general sizes: 1/2 inch for 7/16- and 1/2-inch rope, and 3/4 inch for 5/8- and 3/4-inch rope. It's important that you use the correct size, since the grip is achieved by the cam pushing the rope against the shell. These devices have various breaking strengths but, more importantly, they also have a nominal working capacity. Most of the older versions have a 1,000-lb. capacity, relating to the amount of grip available without damaging the rope. Current catalogs are offering a new Gibbs model that provides a 3,000-lb. working load. This model features a thicker shell with interior protrusions that are curved in shape to increase the grip without inflicting damage to the rope. The basic operation of the Gibbs is to secure it to the rope and clip a carabiner into the hole provided on the cam. Most models are spring-loaded and stay in place, which is an asset. Otherwise, the unit might slip out of your reach.

You can use other mechanical ascenders as well. Do your homework before making any purchasing decisions.

The second type of hard ascender is the handled device. This is familiar to many who have seen mountain climbers ascending rope.

To ascend a rope, you need a minimum of two ascenders; however, three are highly recommended. These devices are sold in pairs, one left-handed and the other right-handed. Ascenders such as these are slipped onto the rope and locked on by engaging the rope between the handled shell and the cam. As you release the cam to take another bite, a safety mechanism prevents the device from disconnecting itself entirely. The operator must remain cognizant because such devices have been known to disconnect anyway. These ascenders are for one-person loads only and must not be used in evolutions where rescue loads occur.

There are two schools of thought regarding the use of hard ascenders. One is to use them anytime you haul or need to ascend. The other is to use them only when the soft type can't do the job. The fear is that the hard type will shear the rope and cause failure or damage the rope during general use. The one thing that most users agree on is that hard ascenders must not be used for belaying or where severe shock loading can occur. Each type of ascender has its attributes and drawbacks.

The attributes of soft ascenders (8 mm cord tied into a prusik on a rope) are:

1. They are readily available and adaptable.
2. They have great strength.
3. They won't disconnect unless the rope or cord fails.
4. The prusik's grip is bidirectional.
5. They're inexpensive.

The drawbacks of soft ascenders are:

1. They can melt and fuse with rope.
2. They can be difficult to tie in inclement weather.
3. Their holding power on icy or muddy ropes is problematic.

The attributes of hard ascenders are:

1. They are easily installed on a rope.
2. Depending on the model, they have superior holding power on icy or muddy ropes.

3. They can be installed and moved into position quickly.

The drawbacks of hard ascenders are:

1. They are more costly than soft ascender cords.
2. They can damage the rope at lower loads.
3. They can shear the rope at lower loads.

Of course, manufacturers are coming out with new products all the time. When purchasing any device, think about its intended application and whether it complies with the standards.

The issue of what type of ascender to use in hauling systems can generate some debate. The bottom line is that, if the system is overstressed or is somehow dynamically loaded, and if the rope grabs must arrest the fall with a rescue load, then severe damage or failure will occur. This is a possibility! Thorough training, knowledge, and safe practices are mandatory. What happens if that equipment bag didn't make it to the scene, or if it fell over the edge? You have to go to your second option, which you'd better know how to execute. Operating the hauling system correctly with communications between the haul team and the ascending load is critical. For instance, suppose 600 pounds of rescue load is being hauled up the side of a cliff. As long as that load is totally dependent on the hauling system and the haul rope is tensioned, then there shouldn't be any opportunity for dynamic loading!

STUDY QUESTIONS

1. According to NFPA 1983 (1995 edition), team carabiners must be those rated for _____ use and have a breaking strength of _____ lbs. or greater.

2. For safety, always load a carabiner along its _____ or _____ axis.

3. Which type of figure 8 descender can best prevent girth-hitching during a tie-low evolution?

4. One advantage of this device is that you can increase or decrease the friction against the rope, and thus more safely control a long descent.

5. Name the three most common uses of pulleys.

Chapter Six

Knots

When I instruct on almost any fire service topic, I talk about my orphans. Orphans are skills that many hate to train on or practice; however, they are also skills that may be put into action at a moment's notice. In our profession, you're expected to perform a host of skills correctly, efficiently, and safely. When it comes to fireground skills, ground laddering and hose stretching are orphans in many companies. When it comes to rescue skills, structural search and rescue is a prime candidate. So is knot tying.

Knots are critical in terms of forming a connection in rope or webbing. Rope wouldn't be an effective tool without them. It's the knots that create a secure bight in a rope for a carabiner to attach to or that allow us to wrap around a pipe to form a tensionless hitch. Knots consist of weaving or interconnecting cordage into bights, loops, or round turns, depending on the particular knot being tied.

The term *bend* is also part of the knot vocabulary and refers to whenever two ropes or pieces of webbing are connected together. Although not technically correct, many use the term *knot* anytime they form a connection in rope or webbing. A popular method to tie two sections of webbing together is with a water knot. This is actually a bend; however, *water knot* is a more common term than water bend.

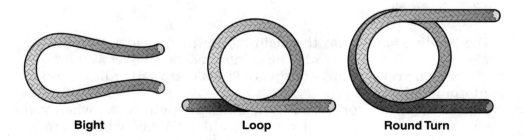

| Bight | Loop | Round Turn |

The firefighter has many knots to choose from but only has to master eight to ten. Learning and maintaining proficiency in them are critical. Knot tying is one of the most basic skills performed, yet shortly after you master the common rescue knots, your ability to tie them diminishes. Individual initiative must make up for any deficit in formal recurrent training. Leaving a short piece of practice rope in the bathroom stall at the station works for many.

SELECTION

There are a few basics to selecting a knot:

1. Use the KISS principle (Keep it simple, stupid).
2. Standardization.
3. It should be easy to tie and untie.
4. It should be easily identifiable by others.
5. You should be able to tie the knot in harsh environments.
6. Choose the knots with the least amount of loss.

You should also be able to tie the common rescue knots in the dark while wearing bulky gloves. Do as much as you can with your gloves on, then take them off for the rest. Obviously, some knots require fingertip dexterity from start to finish—even so, keep your hands protected from high heat and cold wind chills. When working in wind chills of -10°F, gloves aren't optional.

Our need for a variety of knots is predicated on where in a given length of rope a knot is required. Knots tied at an end, as opposed to in the middle, are loaded and stressed differently. This is the reason one knot may be better for one application than another.

Knots in the rescue services area may be grouped into three families:

1. The bowlines.
2. The figure 8s.
3. The hitches.

The bowline family was the mainstay for many generations in the fire service. In the past decade, the eights have taken over as the knots of choice. The third family consists of knots that wrap or lie on top to conform tightly.

One of the major concerns is how much strength is lost when you tie a specific knot. Any knot can violate the 4:1 rule with regard to bending. It's this bending that diminishes the line strength and becomes a necessary evil. We need knots and must choose those that provide the least amount of loss for a given application.

The following knots and bends will be shown step by step without being safetied with a finishing or dressing knot. This is done for reasons of clarity. See the section on the single fisherman for safetying.

THE BOWLINE FAMILY

The Simple Bowline

This knot is still a fireground mainstay because you can tie it around yourself or a stationary object.

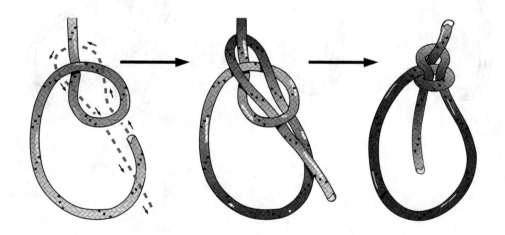

The Double Bowline or Bowline on a Bight

This knot can be used to connect to anchor points or to form the leg straps for a hastily tied rescue harness.

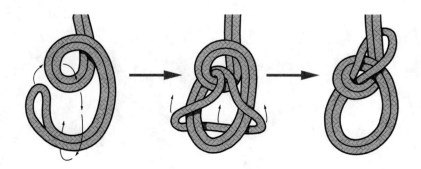

The Triple Bowline

This knot starts out like the double bowline; however, instead of pulling the small bight over the two large loops, you continue to pull the bight until it forms its own loop.

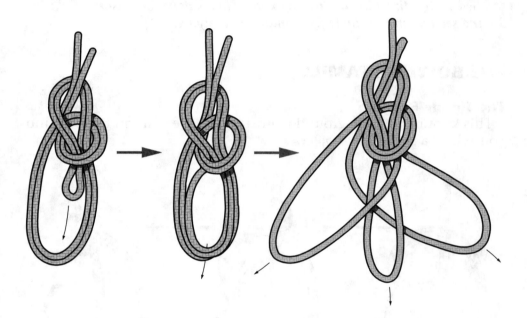

THE FIGURE 8s

The Basic Figure 8 or Figure 8 Stopper Knot

This is the base knot for other knots, such as the figure 8 follow-through. The name stopper knot comes from the practice of tying it in the end of a rappel line before stuffing it into a rope bag. This prevents a rappeller from sliding off the end of the rope.

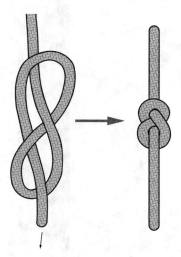

The Figure 8 on a Bight

This is the preferred knot when a secured loop is needed, particularly at the end of a rope. Although not technically correct, many out in the field use the term "figure 8 loop." This knot has lower strength loss than the bowline.

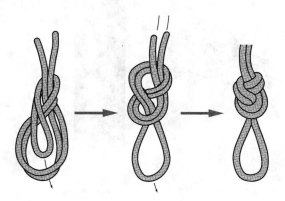

The Figure 8 Follow-Through

This knot allows you to tie around an object such as an anchor point or a harness and have the benefit of the figure 8 on a bight.

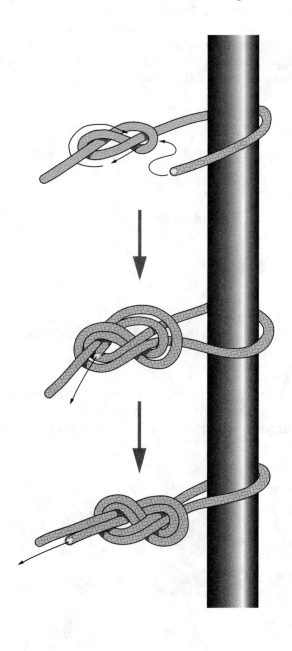

The Figure 8 Bend

This is used to connect two ropes together when a greater length is required. The knot starts out with the basic figure 8 in one rope and the tracing of it with another rope from the opposite direction. This forms a strong yet reversible union. It works best with same-diameter ropes.

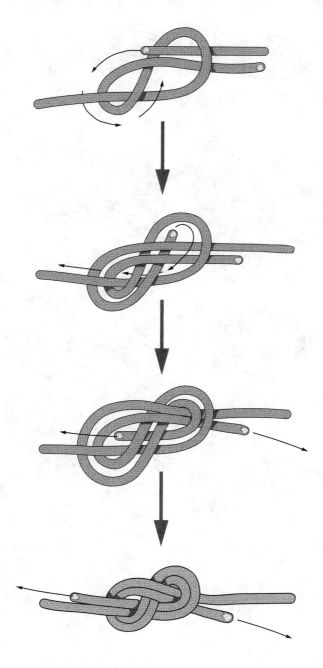

The Double Figure 8 on a Bight

This knot provides two bights (loops) instead of one. It allows for two points of connection.

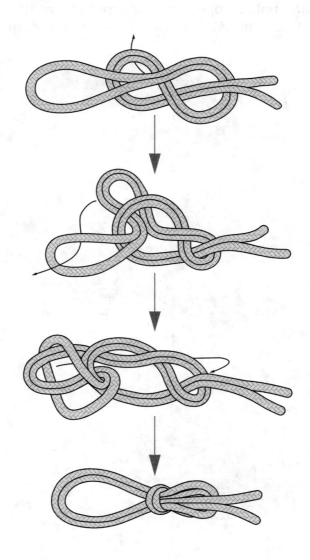

The In-Line Figure 8

This allows you to tie a figure 8 on a bight anywhere and as many as you need throughout the length of the rope. You may have a single- or double-bight tied already. Tie the inline figure 8 as shown.

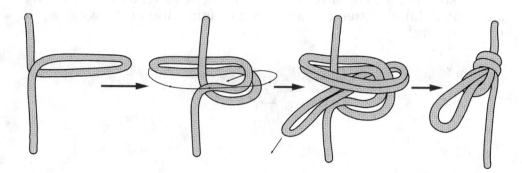

The Butterfly Knot

While this knot isn't part of the 8 family, it's included because of its versatility and strength. You can tie it anywhere along the rope's length, particularly when you need a secure loop and the rope is tied off at both ends. The butterfly is used by many who don't want to tie a figure 8 in a tensioned rope—the pull will come from two directions.

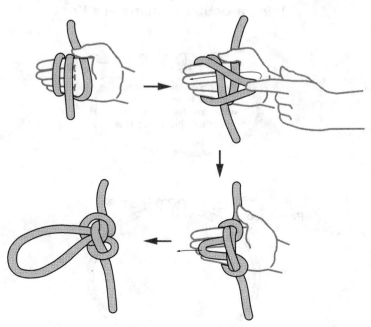

THE HITCH FAMILY

The Girth Hitch

This knot is used either to prevent sliding or to act as a grab. It isn't a life safety knot. Extreme loading and misdirection of stress can set it up for failure.

The Prusik Loop

Select a piece of cordage to form a secure loop. Most in the field use 5/16-inch or 8 mm prusik cord for a 1/2-inch or 12.5 mm rope. Tie the cord into a loop with a double fisherman's bend (knot). This is also known as the grapevine bend (knot).

This is a self-locking and high-strength knot commonly used for a rope grab. Its very composition nullifies the need for backing it up.

Tying a Double Fisherman's Knot

Pulling on the two legs of the loop (arrows) will draw the two single fisherman's knots together.

Prusik loop

Tying a Double-Wrap Prusik Knot

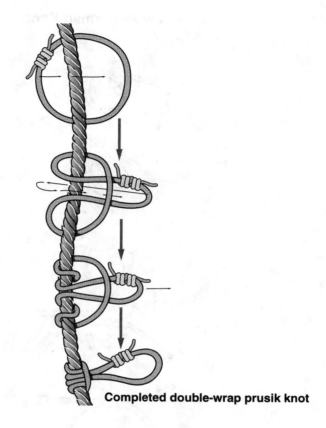

Completed double-wrap prusik knot

Once you form the secure loop, you're ready to wrap it around the rescue rope. Wrapping the loop two or three times around is the norm. The more wraps, the higher the holding power.

The Single Fisherman or Barrel Knot

This knot, also known as the double-overhand bend or grapevine, is used anytime you need to back up a knot.

Author's note: There is some contention in the field regarding backing up a knot. Some feel it isn't necessary and some do. I side with the advocates of backing up. Some of the contention boils down to training issues as each new step is introduced. My feeling is to start every one the same and back up all of them except for a limited few. This makes it second nature. Consider it dressing up a knot and locking down or safetying it.

Single Fisherman Knot

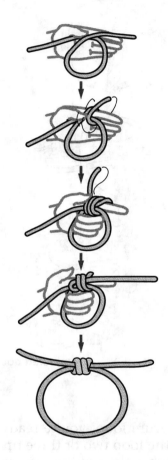

The Tensionless Hitch

This knot is considered a superior method of attachment and is listed as a viable anchoring method. The key here is to find a round
object, such as a pipe, tree, or rock. Rope manufacturers recommend a
4:1 ratio as a minimum for bending; however, an 8:1 or greater ratio
will result in a minimal loss of line strength. A 1/2-inch rope wrapped
around a four-inch pipe results in no loss of strength.

Tensionless Hitch With a Carabiner

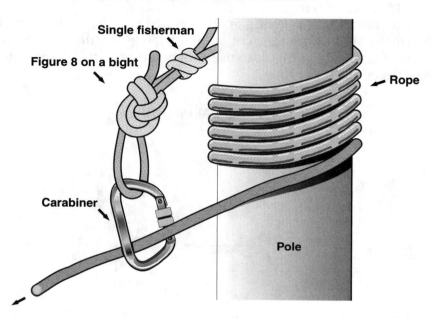

The Load-Releasing Hitch

This is used in both rope and webbing as a release device. It allows you to control the release of a load in such a manner as to prevent shock loading. This and the prusik hitch are used in knot bypassing evolutions on rappel lines. Release knots are beneficial when tensioning is required and, once loaded, would be difficult to get released from the load.

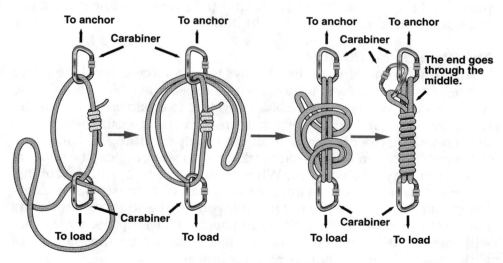

A Load-Releasing Hitch

The Munter Hitch

This is used in a belaying evolution instead of a belay plate. It relies on the friction of the rope contacting itself as it pulls through the cara-biner. This can be difficult to use above one-person loading, and it may be damaging to the rope.

Tying a Munter Hitch

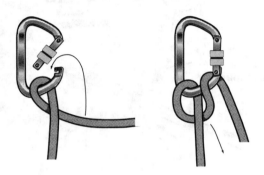

KNOTS FOR WEBBING

This section will look at the few knots available for webbing. Knots for rope and knots for webbing aren't interchangeable in most cases due to design and construction characteristics. Webbing is flat and wide, whereas rope is round and compact. Knots in webbing are designed to take advantage of the available surface area that webbing provides. Much of the knot's strength is predicated on one surface act-ing against another to create a tight bond.

The Water Bend (Knot)

This bend is one of the primary ways to tie a piece of webbing into a secure loop. Once tied and loaded, it's usually permanent. This knot is also for connecting two pieces of webbing together to create one long one. It's made by tying an overhand knot, then retracing it from the opposite direction. Once cinched down, it must be compact, neat, and flat.

Author's note: You'll find that working with webbing can be slippery and that it may creep on you. When working with rope and webbing, you must leave some extra tail for safety. Not only must you tie safety knots on either side of the water knot, you must also leave some extra tail on those knots as well. Some in the field back up the water knot with overhand knots tied on each tail; some back up the water knot with a single fisherman or barrel knot on each end.

A Water Bend

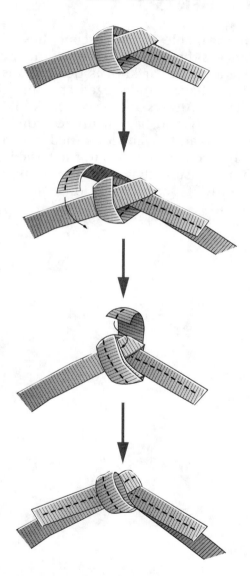

Double Fisherman Bend (Knot)

There are those who tie their webbing slings permanently into secure loops with the double fisherman knots. Once these are tied and loaded, you will find them extremely difficult to undo, yet why would you want to undo them? Once tied correctly, all that is usually needed is some extra tail. The very nature of the knot doesn't require the usual backup.

GENERAL COMMENTS

Many knots are available to the rescuer these days. One problem with this abundance is the lack of data regarding the percentage of line strength loss once a particular knot is tied. This is why those in the rope rescue business have standardized and limited themselves to a group of reliable knots. Always use the correct knot for the job at hand. Tie every knot correctly and back it up accordingly. A correctly tied knot should be neat and easily identified on visual and physical inspection. Dressing up a knot refers to the basic housekeeping of the final product. Making sure that the knot is properly cinched and loaded is important to its security and longevity.

If in doubt, retie it!

STUDY QUESTIONS

1. What is the proper term describing two ropes or sections of webbing tied together?
2. Knots used in the rescue services may be grouped into what three families?
3. What knot is commonly tied into the end of a rappel line for safety?
4. The prusik loop is commonly used for what purpose?
5. To retain line strength, what is the minimum recommended bend ratio to use when wrapping a rope around a stationary object?

Chapter Seven

Anchor Points

In Chapter Two, we discussed breaking down the technical aspects of a rope rescue into three parts: the anchoring system, the access system, and the transportation system. This chapter will begin to illustrate the components needed to construct an anchor system. When you mention this term, many envision a complicated, multifaceted, living and breathing organism. In reality, it can be as simple as one rope wrapped around a pipe. It can also be as complex as a self-equalizing system. In either case, the premise of such a system is that we must attach our human and equipment assets to a substantial object to effect any type of rescue.

There are four basic types of anchor systems:

1. Tensionless.
2. Equalizing.
3. Nonequalizing (load sharing), including single-point anchors.
4. Multiple.

Common to all of them is their need for suitable, substantial anchor points, which are objects that ultimately support the load of people and equipment during a rope rescue. An anchor point becomes part of a larger anchor system consisting of webbing, rope, carabiners, and whatever other hardware is fashioned around it. I purposely refer to an anchor point as an object merely to point out the essential physicality of it. None should be considered truly safe. In reality, you must back up even the most fortresslike object, because it may be a hazard in disguise. Some additional considerations are:

1. What type of loading do you anticipate on the system? Realize that you may set it up for a one-person load, later to discover that it needs to carry two. Do you have the reserve built in?

2. Is the load in line with the intended anchor point(s)?
3. Is the pull directional or nondirectional in nature?
4. Don't construct a system for sport activity if you are going to impose a rescue load on it.

Anchor points fall into three categories:

1. Natural (bushwhacking).
2. Man-made stationary (steel, concrete, or wood structures).
3. Man-made improvised (created on the scene where nothing else is available).

There is no fancy formula or rule of thumb for selecting truly bombproof anchor points. This information is to serve only as a reference. Actual selection must be made only by trained, experienced, qualified, and responsible personnel. An anchor that may hold today may not hold tomorrow, since situations and conditions can change beyond anyone's control at any time.

NATURAL ANCHORS

Natural anchors can be whatever stable objects Mother Nature has provided in the rescue environment, whether animate, such as trees, or inanimate, such as rocks and boulders. Countless resources are available in this realm, but if they aren't present at your particular location, you will have to resort to man-made objects.

MAN-MADE STATIONARY

Man-made stationary anchor points refer to components or construction features that exist prior to a given incident, such as buildings and special structures. Bridges, towers, stanchions, roadway guard rails, and a myriad of other fixed objects can readily serve to support the load of a rope rescue.

MAN-MADE IMPROVISED

It's the improvised anchor points, fashioned on-site, with which we're most concerned. Your operating environment may not already have adequate anchor points, in which case you'll have to create them, whether incorporating preexisting features into the design or not.

Improvised anchor points take advantage of a given location's attributes, particularly some of the nontraditional ones. This can encompass both what the site has to offer and what can be constructed with basic equipment and techniques. Improvising involves the use of common fire service tools and equipment to create suitable anchors. These in themselves require other anchors for stabilization purposes rather than act as primary weight-bearing members. With natural and man-made stationary systems, you tie directly to substantial objects to create a dependable base. In improvising, you construct an anchor point by transmitting the load through a medium, such as a ladder or tripod, to some reliable footing. The newly constructed on-site assembly holds the load while stabilization anchor points secure the operation. The critical point in this process is the newly created anchor point, which in most cases must be centered directly over the load. The overriding question is, how strong are those stabilizing anchor points?

Tripods

Tripods are commercially produced or improvised three-legged devices common to confined space rescue. They support the load of a rescuer or victim to be raised or lowered through a narrow opening. A tripod incorporates a man-rated hand winch that provides the mechanical advantage for hoisting. Some employ a rope system akin to a 4:1 pulley to perform the same function. Follow the manufacturer's guidelines for whatever specific tripod you are using. When shopping for a manufactured model, be aware of the different capacities and options available. You can buy an industrial-grade or a rescue-grade model. Also, read the fine print to see what the capacity is at a particular height. Raising a tripod from a minimum transport length to maximum extension can affect its load capacities. You need to understand the differences between these devices. Matching up a mechanical winch or other hoisting device with the tripod is equally important.

A tripod can also be hastily constructed in the field using substantial timbers or metal piping to form a cohesive load-bearing unit. Its overall integrity will depend on the strength of its components.

The critical dimensions in building a tripod are:

1. The members of improvised wood tripods should be 4 × 4 inches or greater, and be insect- and rot-free.
2. Steel piping should be of three-inch outer dimension and be thick-walled as well as dent- and rust-free (Schedule 40).

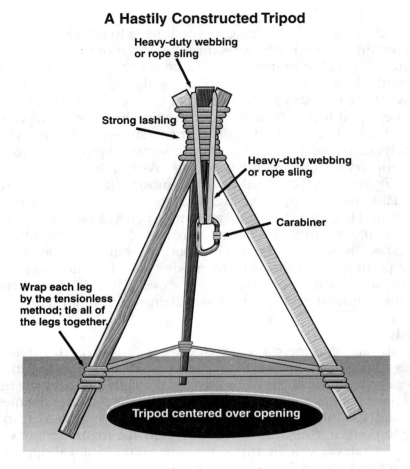

A Hastily Constructed Tripod

Heavy-duty webbing or rope sling

Strong lashing

Heavy-duty webbing or rope sling

Carabiner

Wrap each leg by the tensionless method; tie all of the legs together

Tripod centered over opening

3. Tree limbs and tree trunks from six to ten inches are used in wilderness situations.

You must stabilize the distal legs of the tripod to prevent them from slipping or collapsing. The surface that you build on has a lot to do with maintaining a stable footing. On cement or other hardened surfaces, you must tie all three legs together to prevent a shifting load from popping out a leg. On commercial tripods, the legs are chained together for this very reason.

On soil surfaces, you can drive a steel rod into the ground and tie it directly to the legs. On wooden surfaces, you can nail a rod directly into the floor or box the legs with scrap wood. In any instance, you must tie the legs together as a minimum precaution.

The anchor point is created by looping two webb slings around the axis to form the point of attachment for the mechanical advantage system.

Improvised A-Frame Arrangement

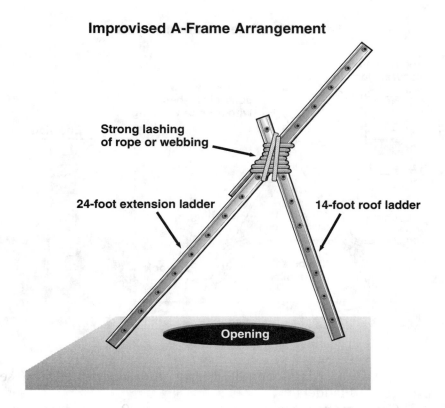

Strong lashing
of rope or webbing

24-foot extension ladder

14-foot roof ladder

Opening

The A Frame

An A frame is an improvised arrangement using two ground ladders secured together at one end and then raised. You must use rated and annually tested NFPA-compliant fire service ground ladders. I'm being specific because, believe it or not, some fire departments use commerical-grade ground ladders because of old habits. The A-frame assembly must be secure at the top and reinforced in the middle to provide stability. The wider the stance, the higher the ladders must be. You don't need ladders that are exactly the same length, as long as the finished A is centered over the opening and you have enough height for the job. Many engine companies carry a standard 14-foot roof ladder and a 24-foot extension ladder as a package. You can extend your 24-foot ladder a couple of rungs and secure the roof ladder to it to create your working anchor point. Such an arrangement is limited to between 400 and 500 lbs. of load. Still, this is a highly versatile method of improvisation. It's a little more stable than the ladder gin and more suited for situations where uneven or differing topography presents a problem.

Critical points with respect to A frames are:

A Completed A-Frame System

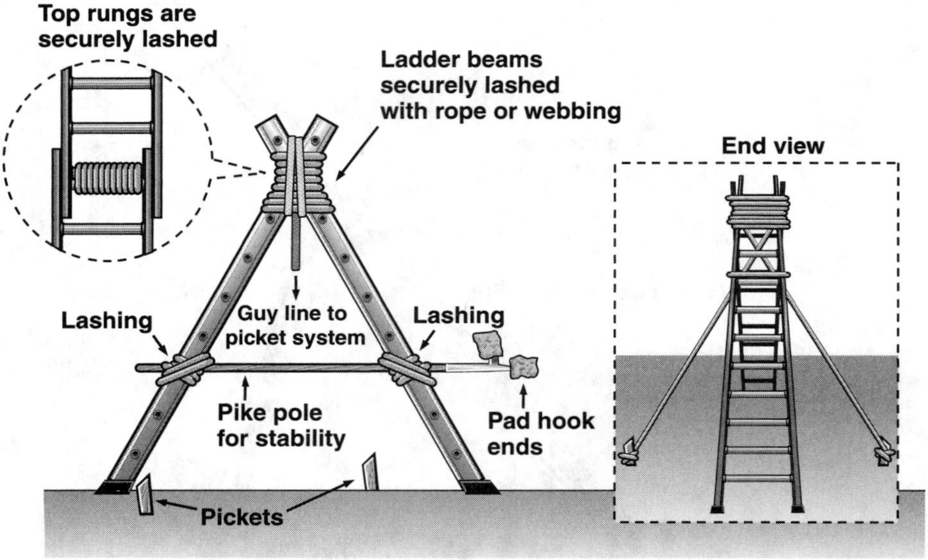

Top rungs are securely lashed

Ladder beams securely lashed with rope or webbing

End view

Lashing

Guy line to picket system

Lashing

Pike pole for stability

Pad hook ends

Pickets

1. Use only NFPA-compliant and annually tested fire service ground ladders.
2. Connect them together so as to form a 70- to 75-degree climbing angle. The ladders are at their strongest in this configuration.
3. Use rope, webbing, and hardware that are rescue grade and able to support the intended loads.
4. Keep all loading concentric within the ladder beams, and avoid any lateral loading. Side loading or unexpected shifting can cause a collapse.
5. Use a belay line in addition to the main rescue line.

The A-frame assembly transmits the load down the beams of the two ladders, and stability is provided by guy lines that run laterally to remote anchor points (see Pickets, below). Tie off the guy lines to anchor points at a distance roughly equal to three times the height of the A frame. Spread the ladder butts about one-third the height at minimum. Tie the two ladders together at the bottom, as well as at the middle, to form a cohesive unit. You can set up a belay line and connect it to the beams and rungs. You can use a static line if you mind it and carefully control it. A firefighter should do this independent of other operations. If any member is in danger of sliding or falling off the grade, secure him to an anchor point with a tether and harness.

A Ladder Gin

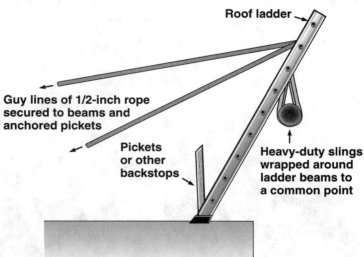

Roof ladder

Guy lines of 1/2-inch rope
secured to beams and
anchored pickets

Pickets
or other
backstops

Heavy-duty slings
wrapped around
ladder beams to
a common point

Gin Pole

A gin pole is a vertical assembly made of metal or wood, supported at four points, that upholds a mechanical advantage system. The fire service's variation is the ladder gin, using a fire service ground ladder. Anchor the ladder into the ground and lean it outward to form an anchor point over an edge or opening. The ladder gin transmits the load from the proximal anchor point to the distal grounding surface. The anchor point, directly over the load, can support 400 to 500 lbs. if properly constructed. Applications for the ladder gin include vertical shafts, wells, pits, vaults, and many other confined spaces. Its critical points are the same as those for A frames. You must support it both vertically and laterally, and it must not take stresses for which it was not designed. Make sure the guy lines are secure, running at 45 degrees and fixed to a ground anchor (see Pickets, below). To create an elevated anchor point, you need a suitable surface backstop for the ladder to butt against. A wall, curb, stump, vehicle, boulder, or the like may suffice. The load will kick out the ladder if it isn't supported properly.

Pickets

Pickets allow you to create reliable anchor points almost anywhere there is soil. A picket is a steel rod or solid hardwood stake. The steel versions are about one inch in diameter and four feet long at minimum. Rebar is popular. The system of picketing relies on a series of

A 3-2-1 Picket System

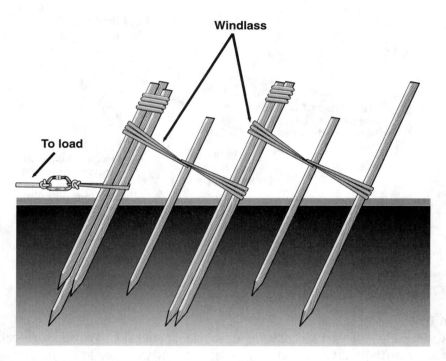

three pickets that are properly tensioned and spaced apart. Such an assembly usually constitutes an anchor point.

The strongest picket arrangement is the 3-2-1 configuration. Although the 1-1-1 is similar in appearance, the 3-2-1 groups a number of pickets together to form a stronger holding system. Drive the pickets into the ground with a sledgehammer, inclined at a 15-degree angle away from the load and sunk about two-thirds of their length. The distance between each picket assembly should be approximately four to five feet. Then connect the pickets with tubular rescue webbing, 7/16-inch static kernmantle, or 1/2-inch nylon laid rope. You need enough cordage to connect each picket assembly and to wrap them five or six times. The lashing begins with a clove hitch and safety tied to the base of the rear picket. After wrapping, secure the cordage to the forward picket assembly with another clove hitch and safety. For tension, insert a short hardwood stake or steel rod between the rope lashing and twist it until the ropes are taught. The correct amount of tension is indicated by the movement of the forward picket assembly. Once the cordage is under proper tension, drive the short windlass stake or rod into the ground. This stake or rod must be long and strong enough to maintain tension throughout the operation.

The weight-holding capabilities of such a system depend on such variables as the type of soil and its moisture content. The holding ability of the 3-2-1 system is between 2,000 and 4,000 lbs. There are smaller picket systems; however, the 3-2-1 is generally considered to be heavy duty and very reliable. The completed assembly comprises the anchor point. The connecting cordage and/or hardware is attached to the first picket assembly facing the load and at the lowest point near the ground.

Exterior Leaning Ladder

This technique uses an NFPA-compliant and annually tested fire service ground ladder to create an anchor point above a window opening. It has been taught in the basic firefighter curriculum for many years and is used for the evacuation of victims from upper floors of buildings.

Lowering a load (one person) is managed by friction created by a rescue rope that is reeved through the rungs. To haul, you can attach a mechanical advantage system to the ladder with webb slings, although lowering will be the appropriate task in most instances.

The critical points of the exterior leaning ladder are:

1. Position the ladder tip above the desired opening of the building.
2. Position the ladder at a 75-degree angle to the building.
3. Properly foot and supervise the ladder.

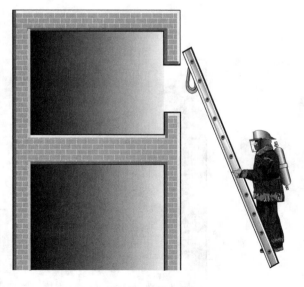

Exterior Leaning Ladder

4. Determine whether you are setting up to lower or haul.
5. Reeve the belay line through the rungs like the lowering line, or establish a topside safety line to a substantial anchor point.
6. Properly package and secure the victim according to his injuries and the urgency of the rescue.
7. Use a harness, stokes, or other suitable, hastily tied rescue knot to transport the victim.
8. Attach a belay line directly to the victim.
9. You may need a tag line to maintain clearance from the building.
10. Have a minimum of two firefighters manage the lowering line.

The above technique is used by companies lacking the proper rope rescue equipment or during extreme emergencies. An alternative to wrapping the rescue rope around the rungs is to wrap a web sling to each beam, connecting them with a carabiner. Then, attach a figure 8 plate as a friction device.

Interior Leaning Ladder

This technique uses an NFPA-compliant and annually tested fire service ground ladder to create an interior anchor point when no other reliable objects are available. This improvised anchor point will allow lowering operations of one-person loads, limited by the length of the rescue rope.

Interior Leaning Ladder

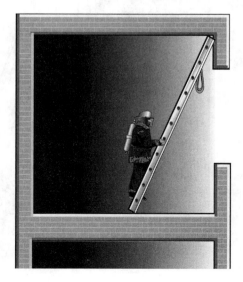

The critical points of interior leaning ladders are:

1. They work well in commercial and industrial environments.
2. A ground ladder that will fit between the ceiling and floor is required. Don't use attic ladders! The small 14-foot extension ladder carried by truck companies will work in tight areas. Remember, this setup is for one-person loads only.
3. Maintain the 75-degree climbing angle.
4. You must properly place and foot the ladder for stability and have members supervise it. You can ladder for low ceilings by removing ceiling tiles or pulling out gypsum board for additional clearance. Many older buildings have several layers of ceilings due to numerous renovations.

Ladder Jib

The ladder jib uses an NFPA-compliant and annually tested fire service ground ladder to form an anchor point based on the jib principle. Place one end of the ladder out a window or over a parapet.

The critical points of a ladder jib are:

1. You must keep at least seven rungs of the ladder behind the edge where the ladder beams rest.
2. You must not extend the newly created anchor point more than one rung beyond the edge where the ladder beams rest.

Ladder Jib

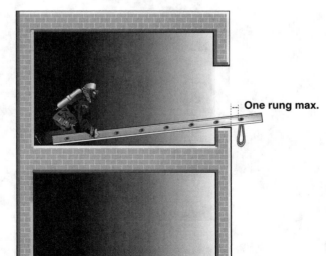

One rung max.

3. You must continuously supervise the activity.
4. A member acting as a counterweight must be in position at least seven rungs behind the edge. This human counterweight kneels on the ladder to avoid slipping. Using more than one member is recommended, as is tying off the ladder's end to a suitable anchor point.
5. Limit the loads to one person.

Cantilever Ladder

This provides the same capabilities as the ladder jib but without having a window or parapet to lean against. The cantilever ladder is used for flat roofs or low openings in industrial situations. Its critical points are the same as those for the ladder jib.

Vehicles

Fire apparatus and other vehicles can be helpful in establishing anchor points if you follow specific criteria. Two obvious hazards accompany using a vehicle: (1) Someone inadvertently moving it, and (2) connecting to a substandard component. Many recommend that you turn off the engine and remove the key from the ignition. There may be times when you have to connect to an operating apparatus, however. It would

Cantilever Ladder

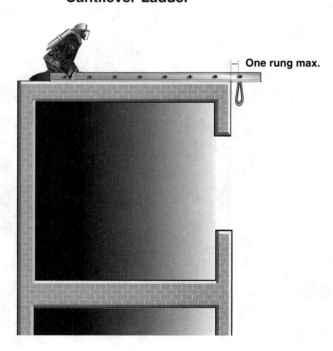

One rung max.

be impossible to turn off an engine while it's pumping or when the aerial is in use. At the very least, place the transmission in park or neutral, as long as this doesn't interfere with the truck's specific function.

Other factors involved in using vehicles are:

1. Consider the vehicle's weight in relation to the load being imposed. Lightweight utility or fast-response vehicles may not be sufficient for low-angle evacuations that include six members and a victim.
2. Environmental conditions such as ice, mud, and rain, along with various grade deviations, can influence the suitability of using a vehicle this way.
3. You must double-chock the chosen vehicle, meaning chocks on two separate tires on the front and rear and with the parking brake set.
4. If the vehicle's engine isn't secured, then someone must be assigned to sit in the driver's seat to guard against any unauthorized use.
5. Supervisors on the ground and in the air must maintain control and clear communications with all involved so that the vehicle isn't moved during the rescue phase.
6. Prospective anchor points on the vehicle must be substantial enough to handle the anticipated loads. Frames, axles, bumpers, and rated tow hooks are likely candidates.

Be careful! Just because it's a tow hook bolted into diamond plate does not mean that it will hold even a one-person load. The following event illustrates this point. In most vehicles, there are two bolt types that you generally don't have to worry about. One is the nader bolt, the other is the one for the seat belt assembly. A recently purchased pumper in my area had a well-hidden deficiency. A firefighter belting in for a call yanked the retracting belt assembly right out of its mount. A subsequent investigation revealed that the assembly hadn't been mounted into the frame with a bolt, but rather, with a sheet metal screw into diamond plate! Don't take anything for granted when lives are on the line.

EMERGENCY MAN-MADE IMPROVISED

This section deals with situations in which there is a likelihood of catastrophic injury or death unless a timely self-rescue is performed. Being trapped in a structure by advancing fire and with no other egress options is the situation that readily comes to mind. When the stairs, escapes, ladders, and adjacent structures are all cut off and no other firefighter can reach you, self-reliance may be the only means of survival. The number of civilians who either jump or throw other family members out the windows of burning buildings is staggering. Reports of such instances are prevalent during the winter months. You probably have seen photos of a firefighter being saved in the nick of time by a hastily thrown ground ladder. The chance of being caught in an involved building is a very real and present danger.

The problem with establishing stationary anchor points in such a circumstance is that there is no time to hunt around for them because the heat and flames are making your location untenable. Your options are to stay and be burned or to get out that window. Getting to the ground safely is the challenge. Lacking other routes, your roof rope or personal escape rope may be your best option. The specific techniques involved in getting down it will be discussed further in the section on rappelling; right now we're concerned with improvising a viable anchor point for the rope.

Attaching to furniture, stockpiles, shelving, and the like is for extreme life-threatening situations only, where if you did nothing you would surely die. There is no guarantee that any of these will support even a one-person load. Still, they may be all that you have available.

You must tie to a substantially large and heavy object. The basic principle is to wrap the rope around that chosen object and tie a secure knot, such as a bowline, or clip the rope to itself with a permanent hooking device, if so equipped. Depending on the weight of the object, pull it as close to the opening as possible to avoiding a dynamic loading situation. As long as the object is larger than the opening and it stops at the frame, then you shouldn't lose your anchor point. Tying to several large and heavy objects follows the multiple anchor principle, as will be covered in Chapter Eight.

Do not employ this technique lightly or without realizing the potential consequences. Even if your anchor holds, many other variables can get you injured or killed. This is your last-ditch effort to try to survive.

Participants in sport or rock climbing routinely improvise their own artificial anchor points. When climbing a rock face, climbers must install points for running protection. These anchor points are spaced out during the ascent and protect climbers during a fall. Man-made improvised anchors are created by installing metal chocks, pitons, friends, and bolts into the rock face, either by fitting them or driving them in. Such devices will not be covered in this book. Using them requires specific techniques, extensive instruction, and practice to master. Snow anchors and deadman holdfasts are also beyond the scope of this text. Instruction in these techniques may prove valuable, however, and can be acquired through wilderness search and rescue groups and mountain climbing schools.

STUDY QUESTIONS

1. Name the four basic types of anchor systems.
2. Tripods, A frames, gin poles, and pickets are all examples of what category of anchor point?
3. When constructing an A frame, the ladders will be at their strongest when they are connected at approximately what climbing angle?
4. The strongest picket arrangement is the _____.
5. List four parts of vehicles that may provide suitable anchor points.

Chapter Eight

Anchor Systems

In Chapter Seven, we covered anchor points, the stationary objects that we use to support a load. In this chapter, we'll cover the actual systems built on them. Anchor systems encompass all of the software, hardware, techniques, and anchor points that form the complete weight-bearing assembly.

The issue of anchors and anchor systems can get convoluted, so let's start from the beginning. I set up Chapter Seven as a specific skill so that you will understand the components necessary to create a good system. To start without good anchor points can set the stage for critical failure. You need at least one bombproof anchor point for any operation. An anchor point must be able to hold the intended load and then some. All too often, the best bombproof anchor isn't where you need it. When you connect to an anchor, use a one- or two-inch webb sling or short rescue rope. Prudent practice dictates using two webb slings as a backup measure if you're connecting to only one anchor point. Be aware that this technique strengthens the connecting point, but if you lose the anchor point, you lose both slings, and the life-supporting anchor system will go with them.

Consequently, you need to back up your anchor point to have built-in redundancy and reserve capacity. A viable anchor system generally uses two or more anchor points to form one substantial point to support the intended load. The critical question to ask is, What intended load is to be imposed on the system? Is it a one-person emergency egress, a two-person rope rescue, or a load even greater? Let's talk about that last case. Rescue loads are considered to be 600 lbs., including both people and equipment. NFPA 1983 presumes one person to be 300 lbs., so a two-person load is 600 lbs. Having said that, there are those who will pack a victim into the stokes and have two litter tenders hanging on to the sides. At minimum, they must be supported by

145

a 1/2-inch static rescue rope backed up by an independent and redundant system. Notwithstanding extenuating circumstances, if you follow the two-person stipulation of NFPA 1983, then 600 lbs. means two persons, not three. I realize that some firefighters have low body weights; however, you must take into account all their clothing and equipment. Yes, firefighters are being protected by lighter-weight fabrics and SCBAs; still, the average member is pushing 250 lbs. when equipped to fight a fire. A technical rescue in an industrial setting may find members in a duty uniform, personal protective equipment, or team equipment. The point is, the fewer people hanging on a rope, the better. In the event of a failure, two casualties are better than three. Such decisions can be sorted out prior to any incident by creating team operating guidelines. These guidelines are tools for the incident and team commanders who must make rapid decisions in the best interests of the victim and members.

Of course, every situation and location is different, even within the same building. This text can't replace hands-on instruction and experience with respect to anchor system construction.

Let's establish a few ground rules:

1. A simple anchor system entails attaching to a single anchor point—the more stable, the better.
2. A complex anchor system entails attaching to two or more simple anchor points.
3. Realize that calling a simple anchor point a "system" may stray from traditional thinking; however, if that's what you go with, then it's a simple system.
4. Do you, in fact, need a simple or a complex system? The situation at hand and all its variables govern this decision.

You can set up an anchor system based on one of four types: the tensionless anchor system, the equalizing anchor system, the nonequalizing (load-sharing) anchor system (including single-point), and the multiple anchor point system. Let's look at the general attributes and drawbacks of each.

TENSIONLESS ANCHOR SYSTEM

This is the simplest of the four options, requiring only one secure anchor point and 1/2-inch static rope. As described above, this anchor uses a tensionless hitch wrapped around an object (see page 123).

Remembering the discussion on pulleys, the 4:1 rule, and the effects of rope bending on the internal fibers, we now expand our parameters to an 8:1 rule. What this hitch provides over the others is minimal loss of rope strength when tied. A 1/2-inch static rescue rope wrapped around a four-inch pipe would retain most of its strength. Obviously, one wrap around isn't sufficient. The suitability of an object depends on its roundness. Any squared edges will diminish the rope's strength. Padding those edges is prudent and may help to reduce bending.

Just how many wraps around the anchor point are appropriate? There's no scientific formula that I'm aware of, but the principle of wrapping takes advantage of the surface area contacted by the rope. Six to ten wraps around industrial piping of six- to eight-inch diameter is reasonable. It depends on the load and other variables, such as weather conditions. When you pull on the standing end and the working end doesn't move (and the securing knot remains untaught), you're pretty much set. Practice and experience give you the edge.

The attributes of a tensionless anchor system are:

1. It is an extremely simple technique.
2. Rope strength loss is minimal.
3. The rope is held without a knot. The finishing knot is for security only and doesn't diminish line strength.

The drawbacks of a tensionless anchor system are:

1. If the main line fails, you have no backup.
2. If the single anchor point fails, your operation fails.
3. Loss of this simple system can lead to catastrophe.

Tensionless Variations

Wrap your primary anchor point with the tensionless hitch. Then, wrap a second bombproof anchor point in close proximity the same way,

Tensionless Backup

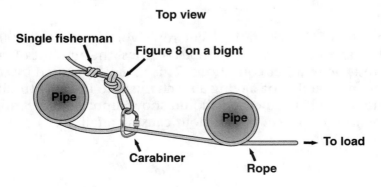

Top view

Single fisherman

Figure 8 on a bight

Pipe

Pipe

To load

Carabiner

Rope

and secure it with a finishing knot. One attribute of this variation is that it backs up the first anchor point. A drawback is that, if the primary rope fails, the backup becomes useless. Also, if the primary anchor point damages the rope and fails, then the backup anchor becomes useless.

Tensionless Anchor with Prusik Hitch

There are two versions of this arrangement, the tensioned prusik hitch and the relaxed prusik hitch (see below). The tensioned prusik hitch entails tying the regular tensionless anchor setup around a secure anchor point. You then tie a three-wrap prusik hitch on the 1/2-inch static rescue rope and attach this to a bombproof anchor.

This variation puts the load onto the prusik hitch and not directly onto the tensionless anchor. This provides a backup and an indicating mechanism. A three-wrap prusik made from 8 mm prusik cord and

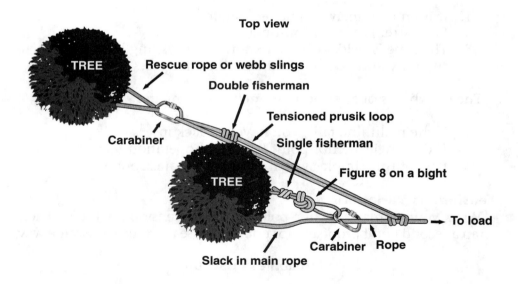

Tensioned Prusik Hitch

attached to a 1/2-inch static rescue rope will slip at about 800 to 1,500 lbs. The variables include any rope coatings, the amount of grip of the cord, and the weather conditions. This variation is used by some who feel uncomfortable not having a mechanism to indicate trouble before a failure occurs. This method should show slippage, indicating that an extreme load occurred, but without causing a failure.

Relaxed Prusik Hitch

Top view

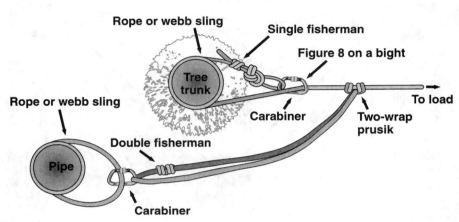

Relaxed Prusik Hitch

This arrangement is set up *almost* the same as the tensioned prusik hitch, except here the load is on the 1/2-inch static rescue rope and the hitch serves as a backup. Obviously, if both anchor points were equal in strength, the rescue rope attachment would be stronger than the prusik cord attachment. Some use this technique to gain redundancy and because the primary anchor point may not be ideally secure. Remember, this is a single-rope technique. This variation is a primary choice for an emergency egress operation or when equipment may be scarce. Also, the tensionless anchor technique can become one of several anchor points in an equalizing system, which we'll cover shortly.

You must understand the ramifications of this and any other anchor system to use it properly. You must also work within the scope and operational guidelines of your team or host department.

The next three anchor systems get to the heart of the critical-angle principle. When using two or more anchor points, it's important to limit the arch, or interior angle of the rope, to less than 90 degrees.

The intent of using more than one anchor point is to divide the load between the anchor points to create one load-bearing assembly. As you know, rope, webbing, or a combination of both is used to form the anchor system. The critical angle affects not only the rope, webbing, and hardware but also the actual anchor points themselves. Increasing the angle between the anchor points increases the stress imposed on them.

Angle Chart

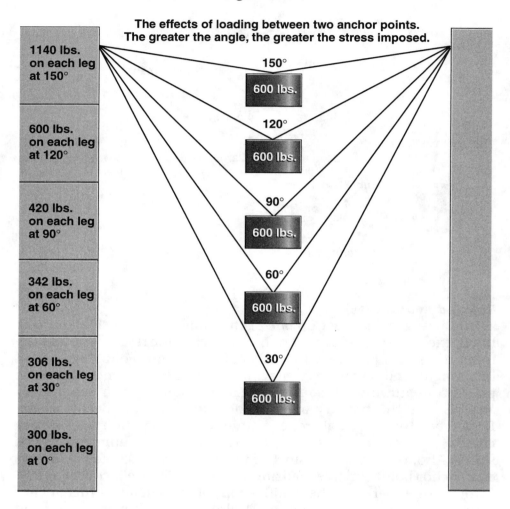

**The effects of loading between two anchor points.
The greater the angle, the greater the stress imposed.**

1140 lbs. on each leg at 150°

600 lbs. on each leg at 120°

420 lbs. on each leg at 90°

342 lbs. on each leg at 60°

306 lbs. on each leg at 30°

300 lbs. on each leg at 0°

150° 600 lbs.
120° 600 lbs.
90° 600 lbs.
60° 600 lbs.
30° 600 lbs.

This chart denotes the effect of loading 600 lbs. between two anchor points. To correct or minimize the stress on your system, you can choose anchor points closer together and/or increase the runout to lower the angle.

For many years, the emphasis has been on the stress imposed on the anchor points with minor emphasis on the other system components. Suppose you connect two bombproof anchors that are 120 degrees apart. You can see by the chart that a 600-lb. load will be imposing 600 lbs. on each anchor, or a total of 1,200. Say that you use one-inch tubular webbing in a self-equalizing arrangement. The safety ratio of 10:1 for webbing (rescue rope is 15:1) will yield a 400-lb. working strength for a new and unused sample. Therefore, even if 1,200 lbs. didn't affect your anchor points, it still exceeded the working strength of the webbing by three times (1,200 : 400 = 3). There's your red flag to trouble!

EQUALIZING ANCHOR SYSTEM

The simplest equalizing anchor system encompasses two anchor points. This forms a nondirectional anchor system that divides the weight as evenly as possible, allowing the new artificial anchor point to be moved where it is needed. The safety twist or notch is there in the event you lose one anchor point. Otherwise, the load-attaching carabiner may slip off the end and you'll lose your system entirely. (The loss of any anchor point can stress the remaining points to such a degree that they, too, can fail due to shock loading.) Many equaliz-

Simple Equalizing System

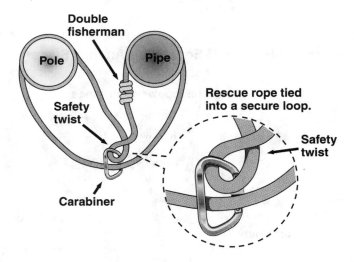

ing anchor systems that support 600-lb. rescue loads use three or four anchor points (see page 152, top). The more points you use, the more time and equipment you will require.

Anchor Plate Equalizing System

This technique is gaining popularity because of its simplicity and versatility. The age-old problem of estimating the size of the anchor harness and knot is eliminated. This technique is based on the creation of the

Webb or Rope Sling Option (Secure Loop)

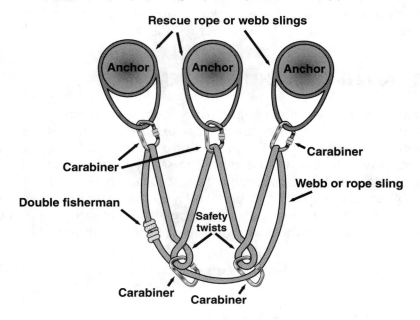

Rescue rope or webb slings

Anchor Anchor Anchor

Carabiner

Carabiner

Webb or rope sling

Double fisherman

Safety twists

Carabiner Carabiner

Anchor-Plate Self-Equalizing Anchor System

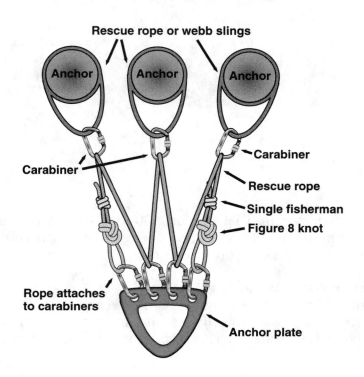

Rescue rope or webb slings

Anchor Anchor Anchor

Carabiner

Carabiner

Rescue rope

Single fisherman

Figure 8 knot

Rope attaches to carabiners

Anchor plate

anchor plate, which is an aluminum plate set up to accept the carabiners from the anchor points to distribute the load and attachment points. It manages the system, keeping everything neat and clean.

Traditional Self-Equalizing Anchor System

This entails tying a 1/2-inch rescue rope into an anchor harness. Short lengths of 25 to 30 feet are useful.

Note: Rescue rope is more beneficial than webbing not only because of its inherent strength, but also for reasons of less friction and easier movement. Theoretically, the load is divided between the objects. In reality, some anchors may bear more of the load than the others for a variety of reasons. Generally speaking, in the field you consider them to

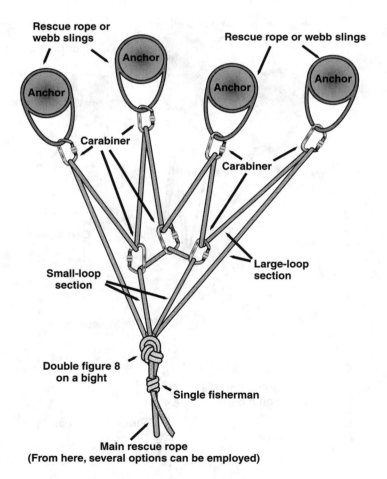

This harness is constructed by tying a double figure 8 on a bight. The trick is to pull one of the bights out much farther than the other. The resultant harness has one very large and one very small loop for the carabiner attachments.

divide the load evenly. If you lose one of your anchor points, the load gets imposed on the other ones. The inherent problem is that the shock load can be enough to cause the remainder to fail. This is why it's important to balance the need for narrow angles between the anchor points and to limit the length of the rope or webbing in the finished loop.

Let's look at the drawback of having extended rope run out in the anchor harness. In the accompanying diagram, a 10-foot run of rope is in place between the anchor point and the connecting hitch. If you lose one anchor point, you'll get an additional 20 feet of rope into the system. This additional slack will accelerate until it runs out. This can cause the system to fail because the length of the drop times the

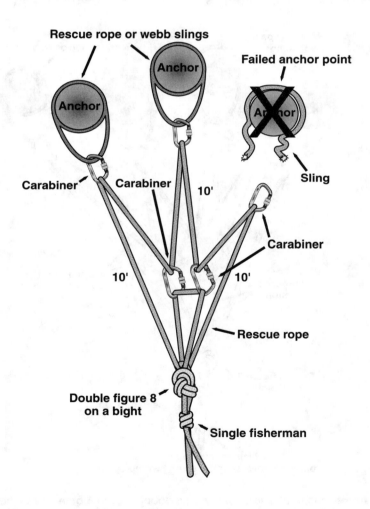

Blown Anchor Point

weight that it was supporting may overtax the other anchor points and blow them out. Using a smaller harness or tether option may better suit your needs. This is also why an independent and redundant system should be in place to catch the load before the 20-foot rope is added to the primary system.

The Tether Option

To eliminate long runouts, you can use a series of tethers run from anchor points to the self-equalizing harness. You can employ a wide

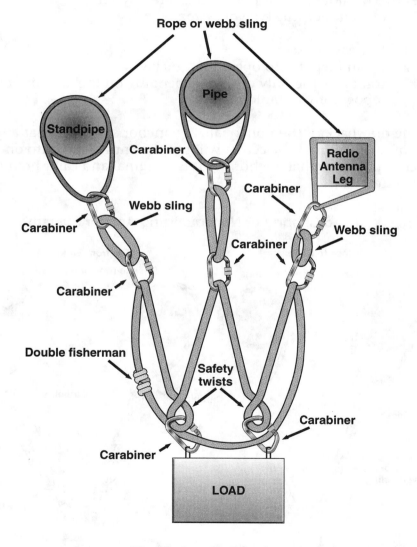

The Tether Option

variety of techniques, software, and anchor point selections. To differentiate this from the previous example, look at the size of the harness. The first one contained 60 to 70 feet of rope, whereas this one is much smaller. The individual tethers allow the connecting to be done from a considerable distance. If you blow out an anchor, the acceleration distance is shorter; hence, the shock will most likely be smaller.

Load-Sharing or Nonequalizing Anchor System

The load-sharing anchor system is a directional anchor system that uses two or more anchor points to create a system.

The attributes of this type of system are:

1. It is easy to set up.
2. It can be put into operation quickly.
3. It may be the only option, depending on the anchor point and topography or building setup.

The drawback of the nonequalizing anchor system is that any shifting of the load from the center will put the entire load onto one of the anchor points. If that anchor point is marginal and can't hold it, then the system fails.

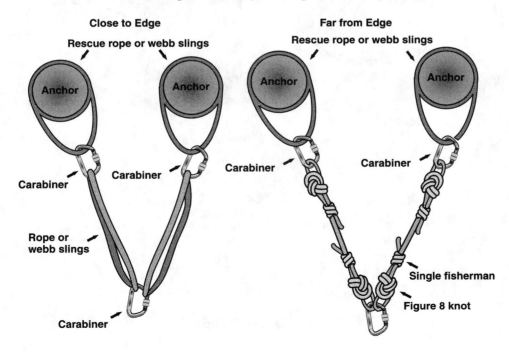

Load-Sharing or Nonequalizing Anchor System

Multiple Anchor

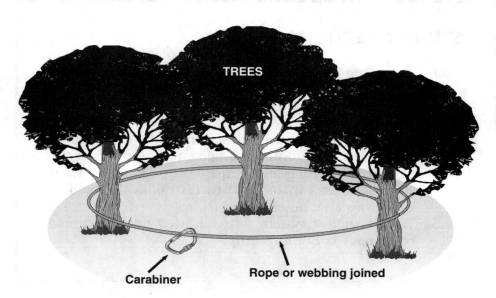

TREES

Carabiner

Rope or webbing joined

MULTIPLE ANCHOR POINT SYSTEM

This technique is only used for extreme emergencies. It could be your only option during a rapidly deteriorating situation when injury or death is a great possibility. This technique entails the firefighter tying off the rescue rope, hooking up to the harness, and egressing the location, most likely the roof. Again, this is for life-and-death situations—just another option in your toolbox.

The attributes of the multiple anchor point system are:

1. It's simple to set up.
2. It requires a rescue rope and a bowline knot and safety.

The drawbacks of the multiple anchor point system are:

1. Without edge protection and proper placement, the chance of rope failure is increased.
2. The 90-degree rule may inadvertently be violated if not enough slack is provided.
3. Correct tying and safetying of the bowline are critical.
4. Loading is limited to one person at a time due to the many variables.

STUDY QUESTIONS

1. According to NFPA 1983, a two-person load is defined as being how heavy?
2. Which anchor system provides the minimal loss of rope strength?
3. When using two or more anchor points, what should be the maximum interior angle (arch) of the rope?
4. In both equalizing and nonequalizing systems, what simple device can be used to distribute the load of numerous attachment points?
5. Which anchor system is to be used under extreme emergency conditions only?

Chapter Nine

Belaying and Running Protection

Belaying simply means securing someone by rope to a substantial anchor. Belaying protects the individual from a fall, since the rope, friction device, and anchor are set up to catch the individual before hitting the ground. The rope is controlled by a belayer, who lets out or takes in line as needed. Any firefighter who needs to climb up, out, or down, or who may have poor handholds and footholds, is a candidate for a fall. Thus, you should set up a belay system ahead of most other protective measures.

The fall matrix consists of three components: (1) the type and qualities of the rope, (2) the length of rope that is out in the system, and (3) the static components of the system, such as the anchor points, carabiners, and webbing.

FALL FACTORS

The fall factor principle is relatively easy to use once you grasp it. To figure it out, divide the length of the fall by the length of the rope that is out in the system.

$$\text{Fall Factor} = \frac{\text{Length of Fall}}{\text{Length of Rope Out in System}}$$

This is one subject that NFPA 1983 doesn't address. It is mentioned in Appendix A for informational purposes only. The primary guiding organization is the UIAA, which sets the standards for climbing equipment. Whenever someone falls and is caught by a protective or arresting system, the resultant impact forces can injure or kill if not properly managed. As they say, "It wasn't the fall that killed him, it was the sudden stop." Tolerances to shock force are figured into equipment and rope to comply with UIAA standards. It has been determined that a 12-kilonewton fall is about the maximum that a human body can sustain without injury.

Let's look at a simplistic illustration of figuring fall factors. These are stand-alone belaying situations without running protection. Fall factors are scaled numerically from less than one to two in this type of evolution. The underlying premise is that you can only fall twice the length of the rope. (This can be greater when clipped into a ladder with a short halyard, which will be covered later.)

Consider the following scenario. A rescuer is anchored to a substan-

Fall Factors

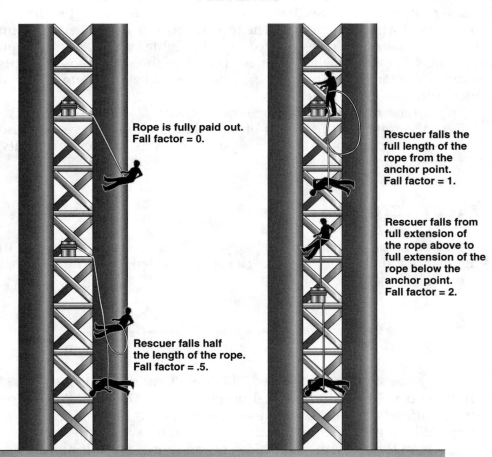

Rope is fully paid out.
Fall factor = 0.

Rescuer falls the full length of the rope from the anchor point.
Fall factor = 1.

Rescuer falls from full extension of the rope above to full extension of the rope below the anchor point.
Fall factor = 2.

Rescuer falls half the length of the rope.
Fall factor = .5.

tial overhead anchor. The rope is pulled taut. Should the rope become shock loaded as the result of a fall, the fall factor would be about zero. In a situation where a substantial anchor is above the rescuer's head, 100 feet of rope is out, and the rescuer is 50 feet below the anchor point, then the fall factor would be 0.5. Falling from the level of the anchor presents a fall factor of 1. If the rescuer climbs the full length of the rope, then falls past the anchor point until the rope shock loads, that is considered a fall factor of 2.

Running protection goes hand in hand with belaying in certain situations and will be covered later in this chapter.

STATIC COMPONENTS

Static components are divided into two categories, equipment and the human body. Equipment components are designed or selected to withstand forces higher than 12 kN, since their utility demands it. A 225-lb. firefighter with equipment is a constant. Outside of shedding some of the equipment load, which is highly unlikely, he will still be 225 lbs. when he falls and be restricted to normal human limitations.

CHOOSING THE BELAY ROPE

Choosing rope for a potential fall factor of greater than 1 means that you should use dynamic kernmantle, which has high energy-absorbing qualities. Vertical climbing and performing a bottomside belay automatically put you in this situation.

Why are we so worried about fall factors, stress, and types of rope? The concern centers around how much stress the human body can withstand in a fall, limited to 12 kN by UIAA standards. In the industrial sector, OSHA Regulation Subpart M, Fall Protection for Workers, covers situations commonly encountered in the workplace. The premise is to eliminate the problem or engineer permanent safeguards where falls can occur. The OSHA regulation covering fall protection limits the maximum arresting force on an employee to 8 kN when wearing a full-body harness. In reality, there is no standard that covers these specific issues when operating in locations without fixed fall protection systems in place. The principles of belaying and its system components provide fall protection where none exist. Our emergency operations occur where there are no safeguards in place and belaying provides the protection to the firefighter. Whenever a fall factor approaches .25 or greater, a dynamic kernmantle rope must be used for belaying evolutions.

What absorbs the shock load? The rope component is the most important in terms of absorbing the shock load of a fall. While some of the equipment items in the static component may yield somewhat when shock loaded, most will do so only above the 12-kN limit. The predominant types of rope being used by the fire service today are the static and dynamic kernmantle. The UIAA standard covers rope used for belaying evolutions in the climbing world. UIAA-compliant ropes are tested to withstand a fall factor 2 situation using a test weight of 176 lbs. (80 kg) and limiting the force to 12 kN. This is the reason using dynamic kernmantle is critical to surviving a fall in lead-climbing situations.

Let's look at some common scenarios. A vast majority of belaying is for horizontal situations such as working on a bridge or vertical situations such as rappelling or being lowered below an anchor point. In the case of a horizontal belay to a rescuer with running protection installed and controlled by a cognizant belayer, a static rope may be suitable. In a vertical descent, such as rappelling with a topside anchor and little slack, any fall should be minimal, and the static rope may also be suitable for this. One of the reasons this topic has been moot is because some teams don't have dynamic ropes in their inventories, and they haven't been trained in running protection techniques. When dynamic rope is discussed, they fear that they must use it in all situations, which isn't true. Still, the need for different sizes, lengths, and types of rope for the multitude of situations that you'll encounter is absolute.

The full issue of fall protection is an entirely separate discipline dealing with workers' safety in the course of their daily duties. The belaying techniques provide a measure of protection to the person operating in a situation without which a fall and injury could occur. The critical balance in belaying is to manage the rope so that enough line is advanced to permit the member to work but not so much as to add to the shock load if a fall occurs. You can see that when the issue of what rope to use comes up, the different concepts and terms can become convoluted. When I talk about belaying, I am generally referring to the safetying of a one-person load. This means that each person requiring a belay should get his own rope, anchor, friction device, and belayer. The associated terminology can have different meanings to different people. Raising and lowering a stokes with a victim and a rescuer requires a main line and a redundant backup system. I don't consider that second rope to be a belay line. It is a system backup, even though the victim and rescuer may be tethered to it. I consider the belay to be a personal and dedicated life safety system. As always, each department or team must discuss these issues and formulate them into their operational plans.

Using dynamic rope may be foreign to some in the fire service; however, you must apply different techniques to different situations. The predominant dynamic rope on the market today is the 11 mm (7/16 inch), which has a breaking strength of 5,000 to 6,800 lbs., depending on the manufacturer. Currently the most common static kernmantle rope is the 12.5 mm (1/2 inch) with a minimum breaking strength in excess of 9,000 lbs. Depending on where your belay station is, climbing above your anchor point can put you into a fall factor 2 situation until you place your first piece of protection. You absolutely and positively don't want to climb above your anchor with a static rope. For anything over a .25 fall factor, you should be using dynamic rope. This runs contrary to the myth that static rope is the only rope suitable for fire service operations.

The lion's share of rescue work is done with static ropes. The problem is that, when you fall any distance, the ultimate shock of arrest is transferred to the rope, the anchor, the harness, the equipment, and your body. You have to worry about not only equipment failure but also the resulting stress on the human body. This is why the emphasis is on dynamic rope in the sport climbing disciplines. For many years, dynamic rope has been instrumental in the successful arrest of one-person falls in vertical climbs.

The guideline to follow, along with manufacturers' recommendations, is to use static rope for most belaying situations where the fall factor is less than .25 and to use dynamic rope for all one-person belays where the fall factor is greater than .25 or whenever you're climbing above your anchor.

SIZE-UP AND SAFETY

Setting up a basic belay system starts with scene size-up. Assessing scene safety involves first determining the safest route to your destination and accessing the start point with a substantial anchor. There are many variables to scene safety, but the primary one to consider is: How tenable is the belay point? You need to find any attendant hazards besides the obvious ones of fire and collapse. Electrical lines, chemicals, gases, machinery, and water can all ruin your day. Those who have dealt with confined space rescue have a good understanding of these

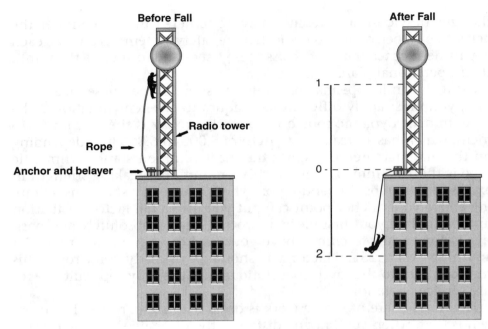

Before Fall

After Fall

Radio tower

Rope

Anchor and belayer

1

0

2

**No Running Protection
Fall factor = 2**

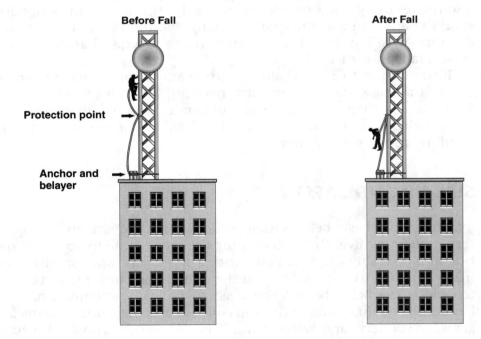

Before Fall

After Fall

Protection point

Anchor and
belayer

**Running Protection in Place
Fall factor less than 2**

hazards. The route to a victim may be safe, but the operating area can be deadly. Keep the start point as close to the victim as possible to make rope runouts short and systems uncomplicated. Substantial anchors in the immediate area must be able to support the load of a belay operation; if not, then an anchor system must be devised. Long-distance or remotely created anchor points are an option, but keep the anchoring principles in mind. Anchors must be able to support the rescuer and victim even if they are using a one-person rope—the victim may jump onto the rescuer and load up the system. Be prepared for such eventualities, and choose your anchor points with care.

Of next concern is ensuring that everyone operating in the danger zone is wearing a harness and is tethered to a substantial anchor. You don't want the mission to be impeded by a personal injury before it even begins.

SETTING UP THE BELAY POINT

There are as many ways as there are situations for doing this. Generally you need an anchor point, rope, a friction device, carabiners, webb slings, edge protection, harnesses, and personal protective equipment. Wrap two webb slings around the anchor points and con-

Belay Plate

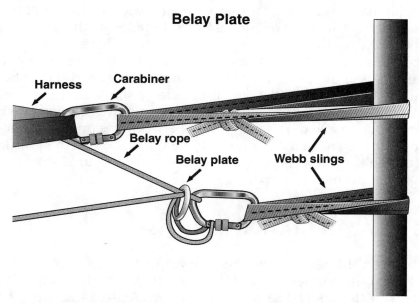

A belay point arrangement using a belay plate. When the belay plate is pulled back, it locks up and prevents any further movement. The belayer must maintain firm control.

nect them to a common point with a locking carabiner, using edge protection as necessary. The first attachment will be for the belay line and must be strongly reinforced. Do the same wrap with a webb sling for the belayer, creating two independent anchor points. The top sling, for the belayer, should be the longer of the two, and the bottom one, for the belay device, should be shorter.

The choice of rope depends on the task at hand. Previously we have discussed choosing rope based on construction and handling characteristics. Now we'll take this a step further. Your choice of a friction device can relate directly to what rope you may use. The primary belay devices are the figure 8, the belay plate, and the brake bar rack. Dynamic rope *will* fit into any of these devices. Static 1/2-inch rope won't fit into the belay plate. The belay plate is designed only for rope up to 7/16-inch diameter. Belay plates are a popular choice of friction device when using dynamic rope. Tandem 8 mm prusiks, figure 8 plates, and brake bar racks are the preferred choices for static rope.

There are many other devices used by teams in different regions and in different countries. When you go above one-person loads, many of these devices have difficulty arresting falls. This is why belaying isn't for multiple-person loads.

Rope selection and belay devices go hand in hand and create yet another area of contention among the experts in the field. Each department and team must make informed decisions as to which techniques and equipment work best for them. A side issue to choosing the right belay device is using a fixed brake in conjunction with it. This entails using prusiks on the rope to act as a backup to the belayer. To many experts, this is a matter of training and safety. The more complicated techniques get, the more recurrent training is required. Let it be said that using prusiks on a belay line for a one-person load has its merits.

If you do elect to use the prusik brake, it's highly recommended that you incorporate a load-releasing hitch between the prusik and anchor point. In the event of a fall while paying out slack and the prusik gets loaded, you'll need a mechanism to release the secured load to continue any movement (See page 241).

THE BELAY OPERATION

With both the rescuer and the belayer in their harnesses, attach a rope first to the rescuer using an appropriate knot. The belayer is then attached to the anchor so as to be independent of the rope. This is where people sometimes become confused. *Belayers don't stop falls by wrapping the rope around themselves.* The belayer simply pulls the rope through the friction device as the rescuer moves away, keeping the rope as taut as possible to eliminate runout, thereby lessening the potential fall distance. When the rescuer comes back toward the belay point, the belayer likewise takes in the slack. The purpose of securing the belayer to an anchor point is to prevent him from falling—since operations usually take place in dangerous areas, an injured belayer could shut down a rescue. The belayer must be independent of the rope and not "body belay," because the stress of arresting a fall could injure or kill him. In the event of a problem, the belayer can tie off the line, leave the belay point, and assist as needed.

In short, belaying is a critical skill that requires thorough mastery of the art. It is a true life safety operation because someone directly depends on your ability to perform the task correctly.

COMMUNICATION

Communication is essential for safety and efficiency. In general, rope rescue lingo is simple and easy to understand. Before throwing a rope bag over a precipice, look to see whether the area is clear, then shout, "Rope." When the rescuer is hooked up to the belay system and is ready to move, he should yell, "On belay." The belayer confirms this by answering, "Belay on." This ensures that everyone is both ready and alert. When coming off the belay for any reason, the rescuer yells, "Off belay," and the belayer answers, "Belay off."

If distance or noise makes direct verbal communication impractical, use portable radios.

PROTECTIVE EQUIPMENT

Team commanders must assess the need for what type and what level of personal protective equipment is necessary. It is essential that you protect yourself against the situation at hand. The ongoing question regarding handling rope is whether rope rescuers should wear gloves. Those who say no back up their position with the argument that operating barehanded provides the rescuer with a constant feel for the rope and thus more control. They also feel that it forestalls a rappeller from taking the extremely fast descents that might otherwise burn and damage the rope. Their basic contention is that standard firefighting gloves are too bulky and dirty from fireground operations to be used on ropes.

Those who advocate gloved operations cite the need to protect the rescuer's hands, especially when operating on steel structures. They believe that fireground work is usually done in full protective gear, including gloves, and that rope rescuers should train this way. For true technical, nonfireground activity, a thin leather glove is usually sufficient, but fireground operations and colder climates necessitate a greater level of protection. On one cold February day, I held a class out on the roof. There wasn't any question as to whether to wear gloves or not!

RUNNING PROTECTION

This technique and belaying complement each other nicely to protect a firefighter in certain situations. For running protection, webb slings are wrapped around substantial components and connected with a locking carabiner to form a secure loop that the belay line will run through. The problem of operating on structures or exaggerated topography is that the belay line can be damaged or entangled by the very object you're operating on. Belaying is called for and used more often than running protection; however, belaying operations can be compromised if running protection isn't used in the appropriate situations. It eliminates many of the rope management problems typically encountered and changes the fall factor facing the lead ascending firefighter. Running protection can be used on many vertical structures, such as radio towers, water towers, and highway billboards. It can also be used on horizontal structures, such as cranes, scaffolding, bridges, catwalks, and pipelines.

Consider the scenario of a rescue from a bridge as depicted. In an attempt to reduce the victim's anxiety and to stabilize the incident, all equipment, nonoperating personnel, and bystanders should be kept remote from the immediate scene of negotiation. By keeping your anchoring point far away, however, you increase the distance that your rope must pay out. As the incorrect setup shows, operating without running protection can result in disaster should a mishap occur. The rescuer could plunge to the terrain below or slam into the structure. It is a common misconception that simply weaving the rope in and around the girders of a bridge is sufficient. When you shock load a rope with the weight of a victim or rescuer, rusty steel edges can cut through it like a knife.

Setting up running protection is relatively simple once a belay point has been established. Before using this technique , you must first meet all of the criteria associated with belays: Prepare the rescuer climbing out

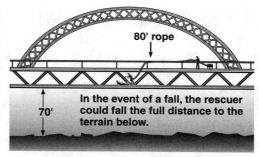

Bridge Rescue—No Running Protection

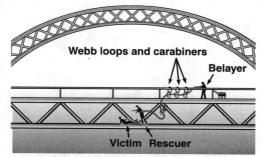

Bridge Rescue—Running Protection in Place

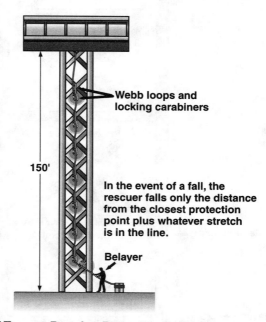

Vertical Tower—Running Protection in Place

or up, establish the belay point and belayer, and have the supervisory clearance. In the meantime, team members can assemble the necessary equipment for the running protection. One-inch tubular webbing tied into secure loops with a locking carabiner snapped to it are the primary components. The loops should be of sufficient size to go around the intended substantial objects. Ladder rungs or beams would require small loops, while structural steel on bridges calls for a larger size.

Zig-Zag Running Protection

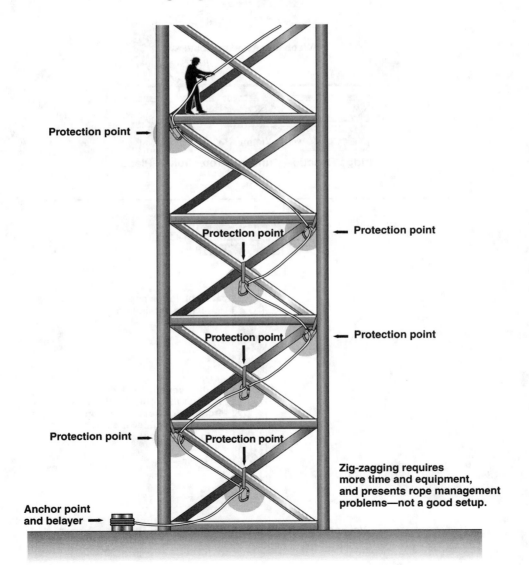

Protection point

Protection point ← Protection point

Protection point ← Protection point

Protection point Protection point

Zig-zagging requires more time and equipment, and presents rope management problems—not a good setup.

Anchor point and belayer →

One of the logistical problems is that most of the slings carried by teams are medium to large in size. Two- to four-foot slings would be sufficient for most applications. On-scene size-up will determine which size you need. If you use larger slings, you may have to wrap them several times to shorten them.

Setting up the system at a rescue begins with tying off to a substantial object to create the belay point. Then, you essentially create artificial anchor points at intervals with webb loops and carabiners.

Running Protection

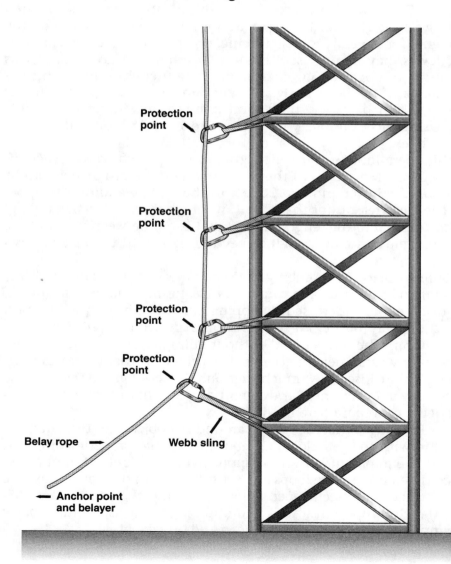

Protection point

Protection point

Protection point

Protection point

Belay rope —

Webb sling

— Anchor point and belayer

There is no hard and fast formula for the distance between each running protection point. Consider the distance to be traveled, how much equipment you need to carry, and the degree of difficulty that the operation presents. With that said, the biggest consideration should be how much impact would result if a fall were to take place. The discussion of fall protection criteria becomes important at this juncture. While the distances that rock climbers cover may at times be considerable, I think that you need to calculate for such distances as they relate to fire service operations.

Take, for example, the scenario of a rope rescue on a 150-foot vertical tower. Place the first protection point just above your head when standing on the ground. Then, place another one every five feet on the lower end for several points. Placing the protection even closer at lower levels may be necessary because, if a fall were to occur at this level and too much rope were available, the firefighter might hit the ground even if the belay is tight. Higher up, at the middle and upper levels, you can place the protection points a maximum of ten feet apart. Using this example, you will require 12 to 15 loop and carabiner assemblies.

In this operation, use dynamic kernmantle rope, which will react differently from static rope. In the event of a fall, the distance would be 10 feet on the lower end and 20 feet in the middle and on the upper end. The basic reason for this is that, when your fall occurs, you'll fall a distance equal to the length of rope that is out between you and the last point, plus an equal distance beyond that until the belay arrests you.

Be sure to place the webbing and carabiners on substantial points and to align and center the carabiners. If questionable points appear, or if distances are increased, wrapping several points to a common connection may be helpful. The running protection point must be able to sustain the impact load without failing. The downside to having too many carabiners or eccentric placement is that it creates drag. You may be surprised how much friction there is when you pull a rope through several carabiners at different angles. That's the reason varying lengths of webbing are needed.

The webb sling should be looped around the object and the carabiner clipped in with the gate down and out to facilitate proper loading and easy use. If equipment is in short supply, you can double 8 mm accessory cord tied into secure loops as a substitute. Most dynamic belays are limited to about 125 feet, depending on the run path of a standard 165-foot (50-meter) rope. On towers and other high objects, multiple belay and running protection operations can and do take place.

Another technique is to establish a belay point for a running protection point to the midlevel and to secure both ends. This is useful if the lead climber needs to be replaced due to exposure or fatigue.

The interconnection between belaying and running protection is important and needs to be clearly understood. Let's look at a vertical tower where the lead climbing firefighter has ascended to the top. The belay line is secured at both ends with running protection in place. Subsequent firefighters will climb the tower superstructure or built-in ladder with the aid of tethers to the belay line. In this evolution, you need to secure your harness to the vertical belay in some manner. There are several ways to do this, all with merits and drawbacks.

Note: You must not overload a structure such as a tower or crane with firefighters. Incident commanders must assess these situations for wind, ice, rain, and type of loads. Structural damage can severely diminish safety margins. Operate on such structures with a minimal number of personnel.

Technique 1

Connect two short webb slings to your harness with a locking carabiner. At the opposite end, connect two more locking carabiners. Clip both to the vertical belay rope, then climb (see page 174).

Once you hit a protection point, unclip one carabiner and clip it to the opposite side of the protection point. Do the same with the second carabiner. With both on the same side, you can proceed up to the next point. Two carabiners mean that at no time can you be detached from the belay line.

The problem with this evolution is that, if a fall were to occur, the firefighter could actually exceed a fall factor of 2 (see page 175). This method is the quickest in terms of advancing up the structure, primarily because the carabiners don't have to be moved along individually. Its overall level of security is for positioning purposes. This fall can exceed the 12-kN limit.

Technique 2

This technique follows the same course of webb tethers and locking carabiners; however, prusik loops used as rope grabs are attached to the vertical belay line. Doing so provides more positive control and prevents you from falling the full distance to the protection point below, as long as the prusiks are gripping correctly. The ascending firefighter must move the prusiks along as he goes and not get ahead of them. The best method is for the prusiks to be advanced between each ascent, the reason being that, if you climb too far above your rope

Vertical Running Protection—Technique 1

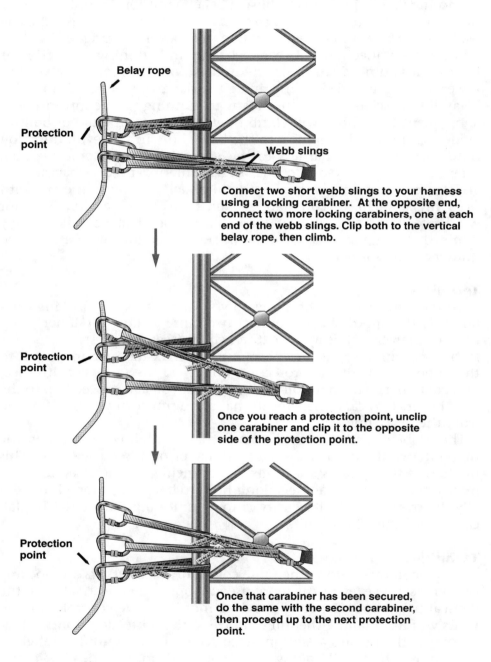

Belay rope

Protection point

Webb slings

Connect two short webb slings to your harness using a locking carabiner. At the opposite end, connect two more locking carabiners, one at each end of the webb slings. Clip both to the vertical belay rope, then climb.

Protection point

Once you reach a protection point, unclip one carabiner and clip it to the opposite side of the protection point.

Protection point

Once that carabiner has been secured, do the same with the second carabiner, then proceed up to the next protection point.

Fall Factor Greater Than 2

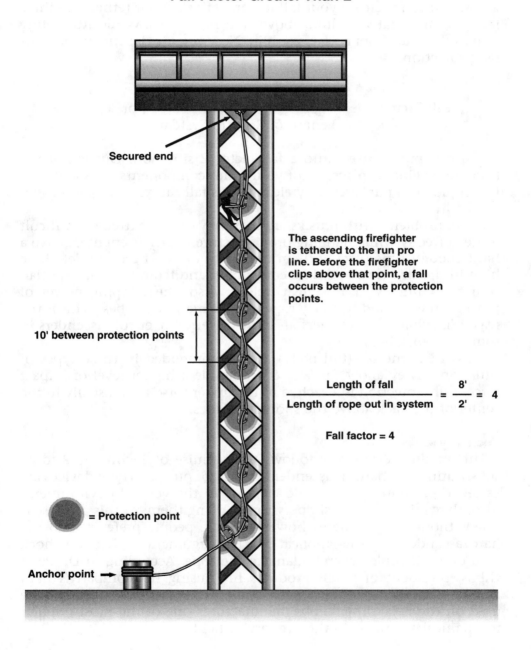

Secured end

The ascending firefighter is tethered to the run pro line. Before the firefighter clips above that point, a fall occurs between the protection points.

10' between protection points

$$\frac{\text{Length of fall}}{\text{Length of rope out in system}} = \frac{8'}{2'} = 4$$

Fall factor = 4

⬤ **= Protection point**

Anchor point →

grabs, you can increase your fall factor. Let's say your tethers are three feet long and that you climb above your rope grabs. While attempting to pull them up from below, you slip. Now you are dependent on your fall protection.

$$\text{Fall Factor} \ = \ \frac{\text{Length of Fall}}{\text{Length of Rope in System}} \ , \text{ or } \ \frac{6}{3} \ = \ 2$$

You can put yourself into a fall factor 2 situation. Outside of the belay line being in place, all of your other components are static and, for all practical purposes, unyielding. This fall can yield impacts higher than the 12-kN limit.

Other problems with prusiks are that they can be extremely difficult to tie correctly while hanging on a structure. Many firefighters have a hard enough time placing them on a rope on level ground, let alone high in the sky. It can be time-consuming and tedious work, especially if the climb is to any great height. Also, the gripping power of prusiks on wet, muddy, or icy ropes can be marginal at best. This is the reason mechanical ascenders are now preferred over soft ascenders in some circumstances.

Class III harnesses (full body) are recommended in these types of situations. They allow the wearer to handle a higher level of impact force. Fall protection is attached to the upper torso area, usually to the center of the back above the shoulder blades.

Technique 3

This method essentially follows the premise of Technique 2, only substituting mechanical ascenders instead of prusiks. These devices are easier to place in service while working in the vertical environment. Also, they grip better on slippery ropes. Using them for this purpose is not without disagreement, however. Some people prefer not to use hard ascenders for this application. The argument is that the shock load of a fall could severely damage the rope. According to Dr. Peter Gibbs, president of Gibbs Products Inc., using his company's hard ascenders in this manner is permissible as long as two ascenders are used on the belay line. Gibbs manufactures one of the most popular mechanical ascenders in the fire service field.

Technique 4

Petzl Manufacturing of France has come up with a new shock-absorbing lanyard called the Zyper™ that can handle a fall factor of 5 with a four- to six-kN yield. As mentioned earlier, traditional fall factors don't normally exceed 2, since you can't fall twice the length of a rope unless it breaks. A unique situation arose in Europe, however, in answer to which this device was created.

The via ferrata is a situation in which individuals climb a fixed ladder along the vertical face of a cliff. A cable runs parallel to the ladder, and climbers hook into that cable for fall protection. This is similar to Technique 1; however, this is cable and not rope, so there is no elasticity. If a fall were to occur, you would fall until you hit the next protection point, where you and your equipment would be shock loaded. This type of fall is possible in running protection evolutions, particularly along the vertical plane. Although rescuers who install their own running protection use dynamic rope instead of cable, they can experience forces that can destroy carabiners, webb loops, anchors, and the human body. This lanyard is designed to arrest an 80-kilogram (176-lb.) person in situations up to a fall factor of 5. The length of fall allowed is five meters divided by the one-meter lanyard, yielding a fall factor of 5. When the fall is arrested, the maximum yield on the

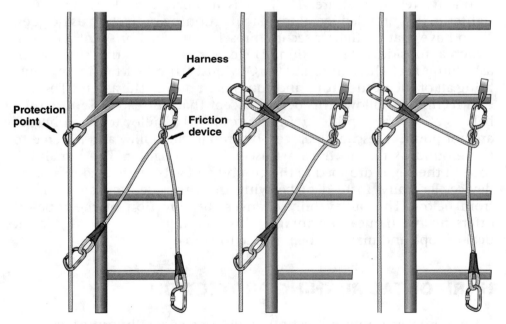

Attach the lanyard to the harness and belay line as shown. When you hit a protection point, clip the free end above it. Once you have clipped the second part of the lanyard above the protection point, you MUST disconnect the lanyard from below the protection point and let it hang freely. You need that second section to hang freely, since it will absorb the shock load by sliding through the friction device in the event of a fall.

Shock-Absorbing Lanyard

climber is to be less than six kN. Remember that it is designed for a 176-lb. person, so a heavier person may want to decrease the distances between each protection point so as to gain a greater margin of safety. This device has made its way through the prestigious CE (Comite European de Normalisation) certification process in Europe. Like all other pieces of equipment, it must be used according to the manufacturer's recommendations.

Technique 4 provides the greatest amount of mobility with a high level of fall protection.

Each of the techniques discussed in the vertical running section have their advantages and disadvantages. Many times, it's the devices that are time-consuming to operate, either because of weather conditions or the relative experience of the user. Having several practiced techniques in your repertoire is prudent, since your preferred technique might be inappropriate for a given situation or the necessary components might not be available. In addition to deciding these issues before performing any evolutions, your team must be highly proficient and versatile.

In case the requisite equipment or skills aren't available, two other secondary techniques are. The first is to belay a lead climbing firefighter to his intended location, placing running protection as he goes. Then have that firefighter secure himself to an anchor point and construct a topside belay station. He can now belay the subsequent ascending firefighter from above. Obviously, the first firefighter must bring all of the needed rope and equipment during the ascent. The second technique follows this design except that, instead of setting up a belay station, the topside firefighter connects a pulley to a bombproof anchor point. A rescue rope, reeved through the pulley and lowered to the ground, is tied to the next ascending firefighter. The remaining rope in the bag is dropped to the ground and gets connected to a belay device. Basically, instead of belaying from topside, you are doing it from below. The pulley only changes the direction of the rope—it offers no mechanical advantage. This technique can require a great deal of rope and may not be possible to employ.

HORIZONTAL RUNNING PROTECTION

Horizontal running protection evolutions generally run at waist or shoulder height. To use the protection in place, you must tether yourself to the system. Connect two webbing loops of equal length to your harness's weight-bearing connecting point. Then, clip two locking carabiners to the opposite end. Clip into the belay line and climb to

Horizontal Running Protection

Reclip the carabiners one at a time to move past horizontal running protection.

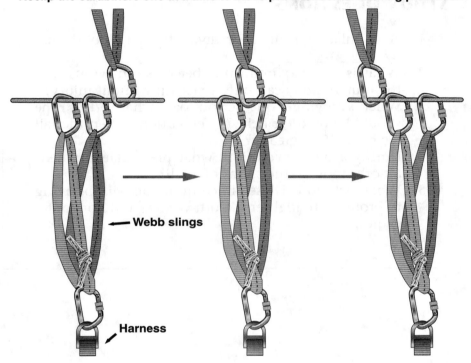

Webb slings

Harness

the next point. Once you hit a protection point, you have to unclip one carabiner and clip it to the opposite side of the point. Do the same with the second carabiner. With both carabiners on the same side, you can proceed forward to the next point. To come back to the belay point, reverse the process. You must always have one of the tethers connected in case of a fall. This is the reason you need two slings and two carabiners. Using the upper lanyard in this case is also suggested.

Using a shock-absorbing lanyard such as the Zyper™ may be more advantageous. It allows rapid movement and additional fall protection in vertical and horizontal situations.

Several options are available to you. You can leave the first lead climber in place and secure both ends, since they can become a belayer to the second lead climber. Otherwise, once the running protection system is in place, firefighters can connect to it and move about as needed, provided the structure can support the weight. Beefing up the running protection line with static kernmantle is possible because the member pulling it through is protected by the secured dynamic rope.

STUDY QUESTIONS

1. To calculate the fall factor, divide the length of the fall by _____.
2. What is the maximum number of kilonewtons that the human body can briefly sustain without injury?
3. When safetying a one-person load, what sort of rope should be used for belaying evolutions when the fall factor is .25 or greater?
4. During a belay evolution, what piece of equipment absorbs most of the shock of a fall?
5. What technique is used in conjunction with belaying to protect a firefighter in both vertical and horizontal evolutions?

Chapter Ten

Access and Packaging

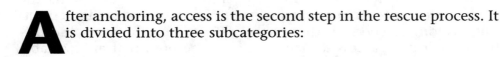After anchoring, access is the second step in the rescue process. It is divided into three subcategories:

1. Making visual, verbal, or direct contact with the victim for rescue.
2. Getting firefighters and rope systems in place to facilitate the rescue.
3. Stabilizing the victim, meaning to package for transport.

This sequence can go in any order, depending on how the operation unfolds. Firefighters may first need to get a rope system in place before being able to make contact with the victim, for example.

Generally, the strategy of taking the higher ground to operate from above allows for simpler tactical objectives and a mission of less complexity. This shouldn't be any secret to anyone, since rope systems require anchor points above the operating area.

GAINING ACCESS ON THE MACRO LEVEL

In Chapter One, we assessed our jurisdiction's liabilities according to four categories: topography, construction, demographics, and in-house capabilities. The first two are germane to our current discussion.

Topography

One segment of a location operating plan is to survey a given environment for access points, both for rope system setup and for reaching the victim. When you look at topographical locations, the issue that presents itself is how to get to the high ground. Many times, roadways provide that advantage. Mountainous locations can require a large logistical effort just to get into position. Reaching that higher ground can be a matter of sheer brute strength if vehicles can't get there. Gaining access to the general theater of operations can expend a tremendous amount of time and resources, and I'm not talking about climbing Mt. McKinley by any means. A moderate hill or ridge in a park may be easily accessible in the summertime. With snow on the ground and strong winds, however, the situation changes tremendously. There are many locations in the biggest city and the smallest town from which typical municipal fire departments might not be able to operate by relying on traditional equipment and apparatus.

Locating the victim via third-party information usually starts the process. Helicopters can be extemely useful in rural and wilderness settings as long as good weather prevails. Unfortunately, many jurisdictions don't have access to helicopters, or the ones that are available aren't configured for rescue operations. In my state, neither the state police nor the fish and game departments provide direct rescue helicopter support. If the need arises, the local agency makes a request to the fish and game department, which in turn contacts the State National Guard Aviation Wing—not exactly a timely event. Regional medical centers that provide air ambulance or medevac service could be your ace in the hole. While they probably wouldn't transport your members, they might pick up your victim from the remote area, thereby increasing his chances of survival. U.S. Coast Guard helicopters may also be available, particularly in the maritime areas. Summertime boating and man-overboard calls in my jurisdiction have brought in several of the Coast Guard's specially equipped helicopters, along with U.S. Navy helicopters from a distant Naval Air Detachment.

One of the best, albeit least known, of the air rescue organizations is the U.S. Air Force Aerospace Rescue and Recovery Service (ARRS). A request received at Scott Air Force Base in Illinois alerts one of the many pararescue (PJ) units located around the world. These units are comprised of helicopters, specially equipped fixed-wing aircraft, or both. They have the capability of flying long distances across the ocean, then parachuting PJs into the sea to assist injured persons. Closer to home, they can reach areas during inclement weather or at night.

With the aid of global positioning satellites and handheld receivers, pinpoint navigational accuracy is available even for ground personnel. Establish a list of these resources and purchase according to your department's potential needs.

Construction

The traditional fire service is more at ease and better equipped to deal with man-made structures than wilderness or other natural locales. Reaching a victim in a building will likely take advantage of all the conventional operations that are the firefighter's stock in trade. Rope rescue operations in man-made structures can take an internal or an external approach. Getting above a victim in an interior environment might be as simple as climbing the stairs to a superior position. External operations might entail climbing the sides of the building, being raised by an aerial ladder, or being lowered by helicopter. Specialized training is available by many of the providers of these more exotic services.

GAINING ACCESS ON THE MICRO LEVEL

The micro level pertains directly to the operating level and can encompass an endless variety of details. This stage of any operation is comprised of many tactical objectives that must be carried out to satisfy the overall mission.

The Fireground

Most likely, victims here will be trapped on the fire floor, the floor above, or the roof. Gaining access and setting up may require a tremendous effort. Whatever positions you establish must be tenable throughout the operation—often a dubious matter when a building is well involved. On the fireground, prudently aggressive tactics are usually ingredients of success.

Man-Made Structures

Access to towers and antennas is gained by unconventional means, whether by climbing an aerial or a built-in ladder or by ascending the superstructure. Belaying and running protection are necessary to provide safety for the firefighters. The lowering maneuvers to get the victim to the ground again can be equally involved. You often have to work out many details at a rapid pace to effect a successful rescue.

Confined Spaces

If, as a minimum, every facility correctly incorporated the required safety systems, and if employers and employees alike abided by 29 CFR 1910.146, *Permit Required Confined Space Regulation,* there would be little use for this growing rescue service. Unfortunately, this is not the case. Spaces are improperly configured, and people cut corners to save time and money. Access in the realm of confined spaces typically implies using the openings that the entrants used and sizing up the anchor points. Depending on the particular industry and facility layout, there may be spots where you can't deploy a tripod and winch. Such spaces need engineered and rated anchor points as part of their design yet often lack them. Consequently, industrial incidents are horror shows in the making. If an entrant goes down, just providing access for rescue personnel and equipment can be a nightmare. A rope tied to the worker in trouble lends a terrible false sense of security and serves as no more than a route to the victim. You must also gain access to the myriad of other hazards in and around the space that you must secure—utilities, chemicals, and machinery, for example. I can attest that the facility lock-out/tag-out program can be extremely flawed or nonexistent, and this may be the primary cause behind many injuries. Rope rescues in industrial settings can involve anchor systems, mechanical advantage systems, and belays at minimum.

PACKAGING

Packaging is done to help prevent further injury, to ensure easier handling of the victim, and to provide a transportation mechanism. Packaging of patients in the emergency medical prehospital setting occurs daily. A typical auto accident patient receives a short board and collar plus a long board and straps. Any fractures will be splinted to provide maximum protection. On a ground level, usually this happens in a rapid, systematic fashion with very few impediments. Packaging a victim in need of rope rescue, however, presents firefighters with numerous ancillary problems. My intent in this section is to make you aware of the limitations of our equipment and to stimulate some thought. There are occasions when maximum prehospital protection can't be given. Maybe the equipment won't fit into the opening, or the operating area is rapidly deteriorating, or the firefighter will be endangered somehow. Many times individuals requiring rope rescue only need transportation to safety, but injuries among such victims are common enough that traditional techniques of stablizing them in the rope context need to be discussed.

Immobilization Devices
The Short Board: Short boarding is a generic term for installing a device on a victim's back to provide rigid support from the cervical to

the coccygeal regions of the spine. In the not-too-distant past, the short board was constructed of plywood and attached by several straps. New technologies have brought about many devices that conform better to the curves of the victim's back. One of the problems with the traditional short board was that it wouldn't fit into some of the newly configured automobile seats. Modern devices are flexible yet still provide ample support once secured to the victim. Such types are commonplace on most ambulances today. The KED™, or Kendrick Extrication Device, is well suited for confined space rescues. Another short board that is available is the OSS II, or Oregon Spine Splint II™. This device was developed to interface with the SKED™ stretcher system, which we'll cover later in this chapter.

The traditional short board still has its place and should be kept around just in case. Of course, using it or any immobilization device requires specialized emergency medical training such as EMT or higher. Follow manufacterers' instructions.

The advantages of short boards are:

1. They provide a high level of spinal immobilization.
2. They help to prevent further injury.
3. They provide for easier handling of the victim.
4. They work well with long boards and litter baskets.
5. They can be applied in many cramped locations. Color-coded straps make attaching them easier in confined environments.

The disadvantages of short boards are:

1. They don't provide a transport mechanism. A basket litter, personal harness, or improvised lift arrangement must be provided.
2. The operating area or egress routes may be too narrow even for these types of devices.
3. Firefighters must be thoroughly trained and competent in their use and must be accustomed to using them in the worst environments.

The Long Board: Long boarding is the generic term for installing a device on which the victim lies in a supine position. The long board provides support from head to toe because the victim is immobilized on the board and held in position. It is generally used in conjunction with a short board for maximum support. The long board is constructed of plywood or heavy-duty plastic and comes with different immobilization features. Wooden types are still popular but are rapidly being replaced by plastic ones. Other considerations include the particular head restraint system and the strapping mechanism. Quick clips that attach to pins in the handholds are advantageous.

The advantages of the long board are:

1. It provides for easier handling of the victim.
2. It helps to prevent further injury by stabilizing the entire body.
3. It is typically used with a short board.
4. The two-board method can facilitate moving a victim through a small opening.

The main disadvantage of the long board is its size. A six-foot board doesn't conform to cramped locations. It is a man-sized device and does not break down. By its very design, it must be rigid to provide support.

Transportation Devices

Stokes Basket: The first tactical stretcher to come to mind is the ever-popular stokes basket, also known as the rescue or litter basket. It has served both the military and the fire service well for many decades. It's constructed of a metal frame of tubular steel shaped to conform to a person and covered with a wire mesh. This stretcher has evolved over

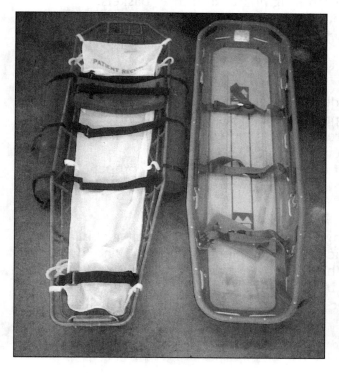

A wire stokes basket with flotation for water operations (left) and a plastic stokes litter.

the years to become lighter in weight and easier to transport. It is designed to be an appropriate means of conveyance for rope rescue, providing the necessary attachment mechanisms for transport. A variety of choices are available from several manufacturers, including:

1. Heavy- and lightweight wire baskets.
2. Heavy-duty plastic baskets with tubular frames.
3. Litters with or without leg dividers.
4. Two-section litters that snap apart for backpack transport.
5. Detachable all-terrain wheel assembly. This is used during long transports and works on the principle of the wheelbarrow.
6. Plastic shields that connect to the frame rails, providing protection to the victim's face and head.
7. Flotation collars for water rescue.
8. Specialty adaptations for towing by a snowmobile or all-terrain vehicle.

SKED Stretcher: The SKED™ is a compact stretcher system constructed of sheet plastic specially designed for confined spaces. This system rolls up into an even more compact arrangement for storing or transporting the SKED and Oregon Spine Splint together. It comes

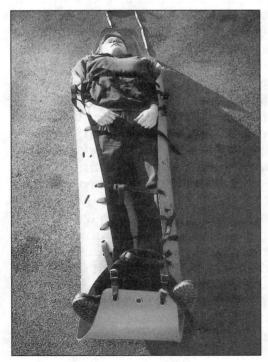

A SKED stretcher.

with several options, including flotation and a rapid deployment system for water rescues.

The general advantages of tactical stretchers are:

1. They provide for easier handling of the victim.
2. They provide the needed mechanisms for point-to-point mobility and connection to the transport system.
3. Heavy-duty plastic models can be used as sleds in the snow.

The disadvantages of tactical stretchers are:

1. They can be difficult to get into position.
2. Loading a victim on a vertical rescue can be difficult, depending on the particular brand.
3. They may not fit through tight corners or small openings.

Whenever using a tactical stretcher, you must follow the manufacturer's recommended procedure for securing the victim. Each of the stretchers covered has its own methods that must be followed.

Placing a harness onto a victim prior to securing him into a tactical stretcher is a common practice. This can done as long as the victim's injuries won't be aggravated. The harness, connected to a separate belay system, is used as a backup in case of stretcher, rope, or knot failure.

Another possible safety precaution is to run some accessory cord or small-diameter rescue rope around the stokes basket rail. Wrap it every few inches until the entire perimeter is covered, then connect it with a double fisherman's knot. Whenever you clip a carabiner around the rail, you must also clip the rope for security.

Attaching lines to a stretcher depends on whether the stretcher is going to be used to haul or lower. In the vertical position, the head area of a stokes is the primary attachment. In one method, the main weight-bearing line is wrapped around the frame rail with a figure 8 follow-through and safety. The secondary backup is attached to a sling with an in-line figure 8, then slackly run down to the tender.

In one method of horizontal attachment, the main weight-bearing line is attached to the four-point connecting harness with an in-line figure 8, and slack is run to the tender. The secondary backup line is also attached to the bridle, and additional tethers are connected to the victim and tender for safety.

The exact method of securing a victim into a stokes stretcher takes on many forms in the field. Whichever is used must be expedient and must also prevent the victim from making any unnecessary move-

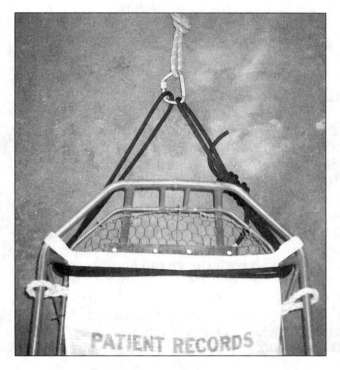

A basic method of attachment for transporting a patient in a vertical postion.

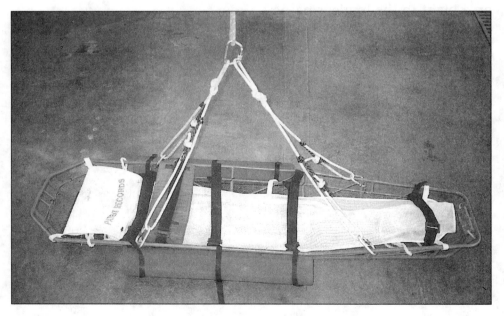

For transporting a patient in a horizontal position. The main point of attachment can consist of large carabiners, a heavy-duty figure 8 plate, an anchor plate, or a steel ring.

ment. The crisscross or shoelace method is often used, as are many adaptations.

TACTICAL STRETCHER MANAGEMENT

Once you are ready to transport a stretcher, you must implement control measures. Install tag lines whenever transporting a tactical stretcher so as to guide it away from edges or obstructions. Other situations require a tender to ride along, either to guide the stretcher away from obstacles or to provide continuous direct patient care. In the vertical position, the tender may straddle the stretcher, walking it around obstacles or guiding it through narrow channels. In the horizontal position, the tender typically hangs midway along the side. This allows for direct face-to-face contact with the victim and easy access to the grab points of the stretcher.

You can tether the firefighter to the stretcher in many ways. Whatever method you use must incorporate a main weight-bearing arrangement and a secondary backup (a belay). One method is to connect the main line to the four-point bridle by an in-line figure 8 knot. Some additional slack is left so that you can attach a safetied figure 8 on a bight to the tender's harness.

Access to an incident area and packaging go hand in hand. Without sufficient access to the victim, complete packaging may not be possible. If there isn't sufficient operating room, then interim transport to a safer region may be necessary to facilitate proper packaging. Rapidly changing environmental factors may also play a role in this decision.

STUDY QUESTIONS

1. Generally, tactical objectives are simpler and missions are less complex when rescuers operate from _____.

2. Which type of immobilization device provides rigid support from the cervical to the coccygeal region of the spine?

3. Name some of the advantages of a long board.

4. Detachable all-terrain wheel assemblies, flotation collars for water rescues, and attachments for towing in the snow are some of the more exotic variations and options of what popular tactical stretcher?

5. When tethering a stretcher tender to a stretcher, you must incorporate both the main weight-bearing system and what other mechanism?

Chapter Eleven

Descending

When most people think about a rope rescue, they generally envision a firefighter swinging from a rope. Although this is an important skill, it is only one of many that you must master. The general technique involved is known as descending. In the rope rescue environment, you generally operate along the vertical plane, where you ascend or descend the rope. Techniques have evolved over the years from the hasty descents and the body rappels to harnesses and various pieces of hardware. Modern methods depend on the rescue rope weaving its way around a friction device to control the descent and act as a brake. Descending is subdivided according to where the brake is located, either at a fixed place in the system or traveling along with the firefighter. A fixed brake allows you to lower the member from a point above. It requires attaching a rope to the firefighter's harness, then attaching it to a friction device connected to a bombproof anchor. The descending member goes over the edge and is lowered by members from above, feeding the rope through the fixed brake and providing all the control. The firefighter is in verbal or visual contact with the edge man, who directs the lowering operation. Some experts view this method as a lowering operation rather than a descending operation since the rescue firefighter provides none of the control.

In a traveling brake system, the friction device rides with the descending firefighter. This technique, known as rappelling, is more recognized and used than lowering. In rappelling, friction created between the rope and the braking device alters the rate of descent and is under the direct control of the rescue firefighter. This is the same technique as is common to sport climbers and the military, among others.

Before a descending evolution can take place, you must have a good understanding of its purpose. You must consider and overrule all other means of reaching a victim before performing a rope rescue, which is an inherently dangerous operation.

There are four general reasons for performing a rope descent:

1. To escape a dangerous position, such as a roof.
2. To reach a victim.
3. To rescue a victim from a dangerous position.
4. To expedite movement from one point to another.

SRT VERSUS DRT

Single-rope technique is referred to as SRT, and double-rope technique is referred to as DRT. Predictably, there are proponents of both types, and a certain amount of debate exists between the camps. The controversy stems from the premise that if something were to happen to the one rope of an SRT, injury or death to the firefighter would almost be a certainty. Double-rope techniques, by virtue of their redundancy, might thus avert a tragedy. Both techniques, I should mention, are generally relegated to one- and two-person loads in a descending operation.

Single-rope techniques originated with the cavers because of the long drops they encounter and their need for simple systems.

The acronym RESCUES serves as a basic checklist for performing a rope descent, whether SRT or DRT:

R = Rope.
E = Evolution.
S = Systems.
C = Consequences.
U = Users.
E = Equipment.
S = Support.

Looking at these criteria more closely:

Rope: Whether static or dynamic; SRT or DRT.
Evolution: What is the task to be performed, and how will it be rigged? Will there be a descent to a rescue? Will a stokes be lowered? Will there be belays and running protection?
Systems: The construction of all the subsystems of the evolution, including anchor systems, access systems, transportation, and belays.

Consequences: This is where you perform the risk-to-benefit analysis of an operation.

Users: The members performing the rescue operation.

Equipment: This refers not only to all the materiel, but also to the logistics involved in performing a given operation.

Support: What parallel operations are needed to support the rescue? Are extra members or hoselines required? Have all the utilities and other hazards been secured? What about technical expertise? Does the situation call for a secondary rescue operation independent of the primary one?

The acronym RESCUES was developed to remind you of the benchmarks required in setting up a rope operation. It serves as a flexible guideline only. When the time comes to operate in the field, many of the training, operating, and equipment issues are already taken care of. The vast majority of fire service descending operations are performed on a two-person rope, usually static. The 1/2-inch static rope setup prebagged with edge protection is preferred.

Under the new standards, harnesses have been broken down into two categories: belts and life safety harnesses. The three types of belts (escape, ladder, and ladder/escape) are meant primarily for self-rescue and positioning of the wearer only. Thus, they aren't suitable for fire service rescue operations. As mentioned in Chapter Three, harnesses are also broken down into three classes: Class I, Class II, and Class III.

The Class II harness is the minimum required for rope rescue operations under both current and previous editions of the standard. The difference now is that the lighter-weight half-body harnesses (one-person) are now categorized as Class I and aren't suitable for two-person loads.

The selection of descending devices and the techniques for using them can inspire spirited debate. The vast majority of descending operations are performed with the traveling brake. Such hardware has improved greatly over the years. The first major device was the carabiner, used singly or in multiple configurations to provide the requisite friction. This worked adequately until the descending or rappel ring was invented. This device went through many stages of development. Commonly referred to as the figure 8 descender, it is now available in many styles. The figure 8 descender has been popular for mainstream fire service rappelling operations for the past two decades. Also popular is the brake bar rack, as discussed in Chapter Five. Many contend that the figure 8 ring is strong and compact enough to carry around at all times, whereas the brake bar rack is large and cumbersome and can be difficult to attach to a rescue rope under less-than-ideal conditions. Pulling one out of your pocket and attaching it *cor-*

rectly in a rapidly deteriorating fire environment is problematic at best. A brake bar rack incorrectly attached can disconnect from the rope. The figure 8 descender requires that you pull a bight of rope through the large hole and pop it over the small one. Once connected to the harness, there is no opportunity for the rope to detach even under the worst circumstances.

COMMON TECHNIQUES

Learning the basics begins on the ground. Master the technique, not the height. All too often, the probie worries about the distance to the ground more than the rope in his hands. Using the acronym RESCUES, a training exercise might break down this way:

Rope: 1/2-inch static kernmantle.
Evolution: One-person rappel.
Systems: A bombproof anchor point (anchor system) and a rappel system (access system).
Consequences: Operating low over level ground—i.e., low risk.
Users: A probie, who may be an experienced firefighter but one who is new to rope rescue skills.
Equipment: Class II harness, appropriate descending device, and protective apparel.
Support: Experienced instructor and attending personnel.

The following is an appropriate way to learn the basics of hooking up to a traveling brake system:

1. Don a Class II harness, and clip on the appropriate hardware.
2. Select and create a bombproof anchor system.
3. Attach the rescue rope to the anchor system.
4. Simulate deploying the rope over an edge by throwing the bag a short distance and having someone maintain a slight pull on it.

Realize that one hand controls your descent while the other guides you. In the grand scheme of things, it doesn't matter which hand does what; however, most rescuers use their dominant hand for braking because it usually has more dexterity and better grip.

With the rescue rope secured and over the simulated edge, it's time to connect the firefighter to the line. If you're right-handed, make sure you're standing on the left side of the rope. This allows you to reeve it

Threading the Figure 8 Descender

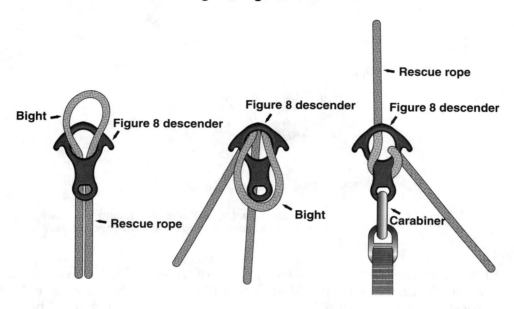

properly through the device and to be in line with the brake hand, a sequence that works well for firefighters in training. Once you master this skill, you'll be able to attach expediently from either side.

5. Hold the device with one hand and, with the other hand, push the bight through the larger hole.
6. Pull some slack through and drop the bight over the smaller hole.
7. With the rope securely in place, connect the figure 8 to the harness with a locking carabiner.

The direction the carabiner gate should be facing is debatable. Because rope techniques and equipment have evolved from the mountaineering and wilderness rescue disciplines, many prefer to keep the gate toward the rappeller. This way, it is less likely to get hung up on an edge and inadvertently open. The problem I've seen with a figure 8 and the gate facing in is that the hardware can shift on contact with an edge, creating severe side loading. Therefore, some personnel attach the carabiner with the gate facing away. The style of the carabiner and its inherent strength may be a factor in this decision. You can eliminate many of these questions simply by getting over the edge. I can't emphasize enough that if you simply keep the rope taught, and if all

Gate Facing In

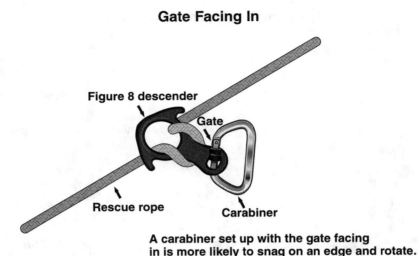

A carabiner set up with the gate facing
in is more likely to snag on an edge and rotate.

of the hardware is clear of the edge, many of these problems disappear.

Another variable you should understand when connecting a figure 8 is topside versus bottomside threading. In the previous example, bottomside threading is used. The premise of those who use topside reeving is that this technique keeps the rope away from the edge better by virtue of being on the opposite side of the device. This may also prevent a girth hitch situation in a figure 8 without ears. I prefer a figure 8 with ears and bottomside threading. Again, your preference will come with expertise. Once you get over the edge, many of these issues go away.

Only use a rescue-grade brake bar rack set up according to the manufacturer's recommendations. Typically this entails a welded eye and six bars. There are some coiled-eye racks that can only hold five bars.

Topside Reeving

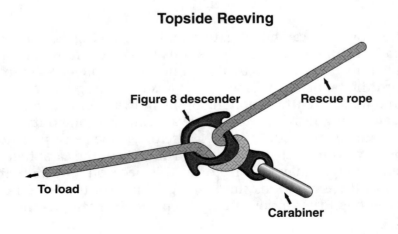

These shouldn't be in a team's inventory. Whatever type you get, you should set them all up the same so as to avoid confusion. There are different diameters, materials, and types of slots available, as well as flat- and twisted-rack versions. The original flat-rack version is still around; however, it's being replaced by the twisted rack primarily because of the D-ring connecting points, which accommodate many of the current harnesses. When used, the rack will turn 90 degrees away, making control of the bars more difficult. To overcome this, some have added a second carabiner to the first to change the plane of the rack. This isn't recommended anymore due to the availability of twisted racks. Also, to add a carabiner is to add one more link that can fail.

The following arrangement is for the older-style brake bar racks. These devices have a 30-year head start on the new NFPA-compliant devices, and they are much more prevalent. The manufacturer will continue to produce the older-style racks.

The first or top bar has a training groove that keeps the rope centered. The second bar usually has a straight slot, and the last four have angled slots. The straight slot on the second bar is to alert you that you didn't connect the device correctly. This points out one of the major drawbacks to this device, particularly on the fireground as opposed to a technical rescue: You can install all of the bars incorrectly, yet it still looks fine. Once you load the device, the bars will pop free from the rack, disconnecting the rack from the rope. Hopefully before you get to that point, the straight slotted bar will drop out, alerting you to the problem.

Brake bar racks can be set up for either a left- or right-handed configuration. Predictably, most are set up for right-handed people. Starting your descent with all of the bars in place is a common practice that provides maximum surface area for friction. Once over the edge, you can spread out the bars or drop the last one. Use caution when dropping out the bars, since you'll lose a portion of the gripping ability.

1. Connect the brake bar rack to your harness with a locking carabiner.
2. Reeve the rescue rope over the top bar, not under it and the rack.
3. Continue to reeve the rescue rope as shown.
4. Examine the threading process for accuracy and alignment.
5. Finally, snug up the bars and load them with your weight before going over the edge.

The closer the bars are to each other and/or the more bars that you use, the more friction you'll have. As a rule of thumb, don't use any fewer

Threading the Brake Bar Rack

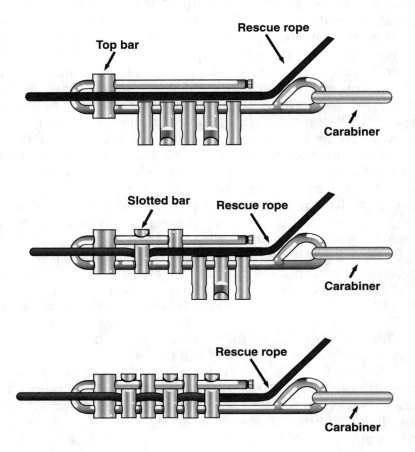

than four bars. Control your descent by moving the bars closer together or spreading them apart with your guide hand, plus steady hand pressure on the brake side as you would use with a figure 8 descender.

BELAYING FOR SAFETY

Belaying is an important safety mechanism, so pay particular attention to it during training. Much of the debate surrounding SRTs versus DRTs traces back to belays. Obviously, all descents are real-life situations; however, some are more controlled than others. Many suggest that actual rescue operations be performed in the securest manner possible, which some maintain is the DRT. The SRT is popular on the fireground, whereas the DRT is popular for training and technical rescues. This is a controversy in the fire service, and your department must decide its own protocols.

In the double-rope technique, attach a separate rescue rope to a bombproof anchor point to support the rappeller in the event of a main line failure.

In the bottomside belay technique, a firefighter on the ground, at the running end of the rope, simply pulls the bottom belay taught, thereby halting the firefighter. The descending member is then free to remove his hands from the rope to grab a victim and be lowered to the ground by the bottomside belayer. This takes some confidence in the bottomside firefighter, who must be competent!

A self-belay entails having a three-wrap prusik loop attached to the rescue rope above the figure 8 device. The free end of the prusik loop is attached to the harness with another locking carabiner. The prusik is moved along by the guide hand as the firefighter descends. The guide hand should maintain just enough pressure to move the prusik and not interfere with the overall process of descending. Note that this can't be done with the brake bar rack because the guide hand is controlling the spacing of the bars and is therefore unavailable to move the prusik. One of the drawbacks to placing the prusik above the figure 8 descender is the constant tension and attention required to manage it. If you don't coordinate it smoothly, the self-belay will take the load and stop the descent. In the grand scheme of things, that's its function; however, it may be difficult to unload the prusik, particularly in a dangerous position.

For some time now, personnel have been trying a newer approach to the self-belay. They have moved the prusik below the figure 8 descender. To facilitate this, they needed to increase the distance of the descender farther away from the harness connection point. A short piece of webbing or daisy chain will do the trick. The advantage to this positioning is that the tension on the prusik is more manageable because the load is more evenly divided between it and the descender. In this case, the prusik is minded by the brake hand rather than the guide hand.

RAPPELLING

For the rest of this chapter, I'll use the term rappelling when describing descending operations.

Rappelling can be performed from most superior to most inferior

points. The degree of difficulty in setting up for a rappel relates directly to where your anchor point is in relation to your point of departure.
The easiest setup is when the anchor point is above the point of departure, allowing for a graceful stepping-back motion to gain proper
placement against a vertical wall, for example.

A problem common to many novice and expert rappellers alike is
getting over the edge. A methodology has been established for this, of
which there are five variations: tie high, tie low, out-a-window
descent, aerial ladder, and tower ladder.

Getting over the edge entails a smooth transition from a stationary
location to a safe descending evolution. This means clearing the structure and moving into an appropriate stance with your descending
device working correctly and safely.

The problems experienced while performing rope rescue evolutions
can be broken down into three categories: education, equipment, and
environment. Educational problems are matters of what techniques to
use and when. Equipment problems relate to using the right tool for
the job, as well as understanding what can go wrong with the equipment and how to get it working again. Environmental problems are
those that can damage or cause the systems to fail as well as injure or
kill those working on them. Getting over the edge involves some of
the educational problems you will encounter when setting up for a
descent. Although the equipment and environmental components are
just as critical, the primary emphasis is to clear those edges correctly
and to get away from the hazardous environment. The firefighter must
be properly instructed and competent in specific descending techniques and equipment.

Getting over the edge can be used in both technical and nontechni-
cal (fireground) evolutions. This methodology was designed with figure 8 descenders in mind. The brake bar rack has its applications; however, whenever there is a rollover situation, the brake bar rack presents
a higher chance of disconnecting the rope than does the figure 8
descender.

Tie High

This arrangement entails using a high-side anchor attachment. The
anchor points must be higher than the descender's point of departure.
A potential anchor on a rooftop would be a substantial radio or TV
antenna or heavy-duty piping that is above your shoulder level. A high
anchor point will help you make a smooth transition from your operating area to the structure's facade or superstructure.

I'll use this first methodology to illustrate the basic premise of rap-

The rope is anchored above—a tie-high arrangement.

The rope is anchored at midlevel—an out-a-window arrangement.

The rope is anchored below—a tie-low arrangement.

pelling. First, stand facing your anchor point with your feet spread about shoulder width apart. Lean back and adjust the slack in the rope to clear the edge. While standing on the balls of your feet with your knees bent, continue to lean back. You're trying to shift your weight and become dependent on the rope. Smoothly and gently let out some slack to transition from the starting point to the vertical plane.

The anchor point for a tie-high arrangement must be above your shoulder level, and it must be substantial.

(Left) No matter what method you use to get over the edge, it is vital that you maintain control as you transition to the vertical plane. (Right) Slowly walk down the wall, stepping around any obstructions.

Eventually you will lean back far enough to reach the point of no return. This is where gravity takes over. Without help, you won't be coming back over that edge. If you lean too far and don't bend your knees, you'll flip over backward and end up inverted. If you don't lean enough, you may slide down the wall, hitting your knees and face. No matter what problem occurs, once you're over that edge, you need to focus on the task at hand and regain your stance. You may bang a few body parts, not to mention your ego, but the job is to overcome that and continue the evolution. Fine-tune your stance as you come across any obstructions or protrusions. Slowly walk down the wall, stepping over or around windows, fire escapes, and wires.

Tie Low

This arrangement denotes a low-side anchor attachment. The anchor point must be lower than the descender's point of departure. It may be as low as the floor level. A common situation would be a parapet along a rooftop. This can be approached in two ways: Either stand on the parapet or roll over it. Standing on the wall and performing a rapidly controlled drop isn't for everyone. You'll drop the distance from your harness attachment to the edge you were standing on, then an equal distance until the ropes takes your full load. If not done with some finesse, this five- or six-foot drop could prove dangerous. The tie-low technique facilitates a smooth and gentle rollover maneuver that avoids severe dynamic loading of the rope system and any padded edges.

This method is used when your anchor points are level to or lower than the point of departure. Hook up your figure 8 descender correctly. In the beginning of this chapter, we discussed right-handers standing on the left side of the rope. This is a good format to enforce this

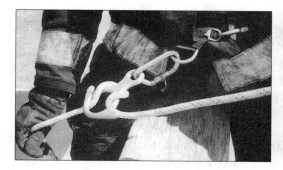

(Left) Before going over the wall, pull enough slack in the line so that the hardware will clear the edge. (Right) Place your guide hand on top of the rope so that it doesn't get caught underneath. (Bottom) Rolling over the wall is probably the easiest way to find the correct stance when using a figure 8 descender.

rule for novices. Typically, once the rescue rope is anchored and dropped over the edge, you stand parallel to it and hook up. Your guide hand will point toward the anchor and your brake hand toward the edge.

Make sure the edge is intact, secure, and protected. Straddle it and pull enough slack in the line for the hardware to clear it. Now, prepare to roll over the wall. The guide hand must lie on top of the rope so that it doesn't get caught, inadvertently providing a pivot point. The brake hand punches the wall and holds its grip. The guide hand is still on top of the rope. If the rope were to stay on your hand, you could be trapped by your own weight and be left hanging. Roll over the wall to regain a correct stance. Once over, check to ensure that the rope is threaded correctly, then proceed to the ground.

Prior to developing these methodologies, the predominant way to clear the edge was simply to walk off it. Doing so with a tie-high arrangement doesn't pose much of a problem; however, the tie low presents several. Walking off a parapet can be difficult if the firefighter has not practiced that skill. Rolling off with the tie-low technique tends to be less stressful on the wall, rope system, and firefighter. When using a brake bar rack, walking off may be more suitable. This may influence your operating guidelines and what equipment you select.

Out-a-Window Descent

This arrangement follows either the tie-high or tie-low arrangement in terms of the anchor point's location, with the additional problem of clearing a window frame. The out-a-window technique requires an opening large enough for a firefighter to exit. This would include a window that can be physically opened or taken out. Most residential or apartment occupancies would allow this. Remove any drapes or curtains, and clear the immediate area around the window. If the window is extremely large, then opening the bottom sash and getting rid of the screen may be all that is necessary.

Setting up for a descent out of a window depends on the type of occupancy and the kinds of anchor points available. Going out of a window is usually to escape an untenable situation or to reach someone in a window below. It is best to take out the entire window so that the firefighter can straddle the sill and eliminate the hazard of glass. Taking out the window entirely means just that—*entirely*. Many of the replacement windows currently in use tilt in and pop out of their

(Left) Taking out the entire window will simplify egress, particularly when you are laden with full equipment. (Top right) You must incorporate edge protection when going out a window. (Bottom left) Pivoting on your outside foot will help you assume the correct stance.

frames and so can be put to one side. If not, a little persuasion against a stubborn one will usually suffice. In any case, the work of an aggressive truckie may be the ultimate answer. Your prebagged rescue rope must have an edge protector attached and ready to go. You can use readily available materials on the scene such as bedding, carpeting, or draperies in an emergency. Select your anchor points and hook up your rescue rope, then toss the rope bag out the window.

After you have hooked up to the line, straddle the windowsill and give enough slack so that the figure 8 will clear the edge. It is important to move as close as you can to the frame in front of you. Egressing with an SCBA, you'll need all the room you can get to pivot out that window. Pull the figure 8 clear of the frame. The inside guide hand holds on to the interior wall, and the outside brake hand punches the wall and maintains a firm grip. Now, pivot on your outside foot to bring your entire body out. Check quickly to ensure that the rope is threaded correctly. Get into a proper stance and proceed with the descent.

Aerial Ladder

The use of an aerial ladder is indicative of confined space, below-grade, or collapse operations that necessitate a firefighter being inserted. Reliable, conventional anchoring may not exist. Of course, you must only use aerial apparatus rated for the loads being applied and never exceed manufacturers' limitations. The aerial must be maintained and tested per NFPA 1914, *Standard for Testing Fire Department Aerial Devices*.

In some situations, SCBA may be required. If so, you must include it as a factor in any relevant decision making. If you are forced to wear SCBA, you must don it correctly, with all of the straps applied snugly. If the waist straps aren't on and pulled snug, injury could result.

Both the aerial ladder method and the tower ladder method (see below) can be considered special operations skills. These rapid insertion techniques allow a firefighter to reach areas that were formerly inaccessible by conventional means. The exact method to be shown here is a single-rope technique. If you use a belay system, careful placement of the belay line is important. If using one creates rather than eliminates a hazard, consider disregarding it altogether.

Keeping in mind the stresses imposed and the limitations of the apparatus, select and establish your anchor system. This can consist of rungs, rails, or turntable superstructure components. One-person loading is prudent.

Next, position yourself near the ladder tip. Hook up your rescue rope correctly and connect it to the harness. The rope will run between your

(Left) When using an aerial ladder, you may use the rungs and rails to establish an anchor point. (Top) The guide hand holds on to the last rung, and the brake hand holds the standing end of the rope. (Right) When going over the edge, use your brake hand to keep the rope taut.

legs. Pull enough slack so that the figure 8 clears the edge. The guide hand holds on to the last rung, and the brake hand holds the standing end of the rope. Note that the guide hand must be positioned *on top* of the rope so that it doesn't get caught when you roll over the rung. If you grip the rope and rung together, you shouldn't experience a problem.

After a safety check and clearance from a supervisor and belayer (if present), the rollover can begin. Go over the last rung very slowly,

Going over the edge of an aerial ladder.

maintaining your grip and control. Your brake hand at this point is over the edge, maintaining tautness. Once in the correct vertical position, ensure that the figure 8 is threaded correctly. Release your guide hand from the rung and place it on the rope above the figure 8, then proceed with the descent.

Tower Ladder

Using a tower ladder provides the best operating parameters when aerial apparatus is required. Heavy load capacities at low angles and full extension are some of a tower ladder's attributes. The method described here is a second version of the rapid insertion techniques available. Using a belay system that is managed from the basket with a device that's anchored below can be difficult to do. It is a premise of this arrangement that the firefighter carries a rescue rope prepacked in a rope bag. Once the firefighter is over the drop zone, he drops the bag and then descends. To eliminate the topside belay and still have a double-rope system, you can set up the rappel system described below.

Doubling up two 1/2-inch static rescue ropes, bagging them, and then descending can be a little cumbersome. There is a lot of 11 mm (7/16-inch) static rescue rope on the market that can serve you well in this evolution. These ropes have a breaking strength of around 6,500 lbs., and they are considered one-person ropes under the standard. You can pack these two ropes together in the same bag. When you're done, you'll have a doubled rappel line.

First, tie a double figure 8 in the running end and dress it up. Then, tie a second double figure 8 about four feet away. Clip locking carabiners into the rated lift points of the tower, and connect the first figure 8. Then, do the same with the second one.

A doubled rappel line anchored to a tower ladder.

(Top) Once you are connected, tie or lock off the descender. (Middle left) Clip the rope bag to your harness. (Bottom left) When positioning the firefighter for the drop, consider the wind and any elevated hazards. (Right) After you have deployed the rope bag, execute the rappel as you would in any other free-hanging evolution.

Connect the rescue rope to the descender, and connect to the harness with a locking carabiner. Tie off or lock off the descender. This is to relieve the firefighter from maintaining a firm grip during the entire process of moving from the ground to the drop zone.

Next, clip the prepacked rope bag to your harness. This eliminates the danger of the rope's contacting overhead obstructions. Slowly lift the firefighter up and position him over the intended drop zone. Remember to make allowances for wind conditions and elevated hazards, and don't drop the rope bag on or near the victims. After he throws the bag, the firefighter unlocks the rope lockup and rappels to the ground as in any other evolution.

Do not use aerial apparatus without proper forethought. Incorporating it in an operation has its risks and benefits, and it requires strict control and coordination, involving both command and operations. The apparatus must be placed properly, and any hazards on the scene must be dealt with prior to using these techniques.

WINDOW RESCUES

Having touched on window evolutions, let's discuss rescuing victims from them. The two general methods involve operating with and without a harness. The without-harness category usually implies trapped civilians. With-harness rescues usually imply firefighters who need to get picked off by another member. Our discussion will be limited to victims who are conscious and able to get into position for the grab. Unconscious individuals require some sort of interior assistance and packaging.

Communication
First and foremost, establish verbal communication with the victim. Tell the victim to stay where he is and what you're going to do. While the rescue system is being set up, assign a firefighter to maintain contact with the victim, staffing permitting. This is important because time is critical, and reestablishing contact is time-consuming. Communication may be difficult due to a language barrier. If such is the case, try to get a firefighter in place who can speak the same language. Ultimately, this will make for a safer operation.

Double-Wrapping a Figure 8 Descender
When a firefighter rappels with a second person, the increased load can make controlling the descent difficult. Consequently, more fric-

Double-Wrapped Figure 8 Descender

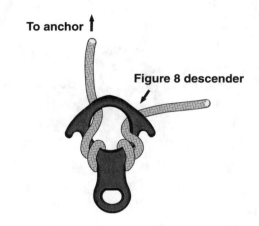

tion is required. To provide more contact between the rope and the figure 8 device, you must double-wrap it. Start out as you would for a single wrap, only this time pull through a little extra slack. Then, wrap it a second time. Double-wrapping a 1/2-inch rope may create too much friction for the single-person load of a firefighter during the initial phase of the descent. It may require feeding the rope into the figure 8. Once the second person is on board, the descent should go more quickly. Practice is vital to developing the skills necessary to making safe descents with a two-person load.

No-Harness Rescue

As mentioned above, rescuing a victim without a harness usually implies a civilian rescue. Remember that if the rope is dropped from above, it must be out of the victim's pathway. This is important so that the victim doesn't hold up the operation or decide to jump for the rope. Also, when rappelling down to the window, keep your brake hand away from the victim. Stop above the window and tell the victim what you are going to do. The reason for keeping your brake hand on the outside is fairly obvious: It prevents the victim from grabbing it and stripping you of control. In an urban environment, you usually have the benefit of a bottomside belayer who can control the operation if a problem arises. Otherwise, regaining control might be difficult if not impossible.

Tying or Locking Off

In making such a rescue, you must be able to stop and secure your descent so that you can remove your hands from the rope evolution.

Tying Off the 8 Plate

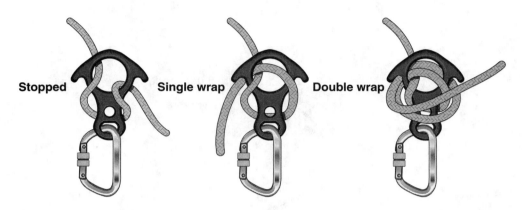

Stopped   **Single wrap**   **Double wrap**

The technique for this involves wedging the rope between itself and the figure 8. This requires a little dexterity and practice. From a normal braking position, move your guide hand down and around the carabiner and the figure 8. This prevents any twisting action and makes the process more stable. Never forget that you're suspended in the air! Don't let your brake hand go, since your life depends on its grip. Maintain about a hand's width above the figure 8 with no downward slippage. In one quick motion, cross over from the brake side to the guide side. In a right-handed example, it's from right to left. You are pulling the rope downward between the vertical portion and the figure 8. Keep steady pressure on it until you pop it into a secure position.

This requires practice and, at times, patience. When done correctly, you will be prevented from moving downward. This technique works extremely well against a wall or other surface. With additional practice, it can be done in free air. It's easier when trying to pop the rope in place to pull the figure 8 inward and then step up slightly. One wrap will hold you in place; however, two are required as a safeguard. The second wrap is done in the same way as the first. Safety the wraps with an overhand knot. When untying or unlocking, simply reverse the previous steps, starting with the safety knot. Then, grab hold of the rope and unsnap the wraps one at a time. Remember that you must maintain control of the standing part of the rope, the brake side.

With a brake bar rack, come to a complete stop, then pull the rope up toward the anchor point and the top of the rack. While maintaining a firm grip and control with the brake hand, wrap the rope between the top bar and rope. Wrap it twice at minimum from bottom to top. Safety the wraps with an overhand knot, then inspect them for security. Reverse these steps to untie.

Tying Off the Brake Bar Rack

Avoid pinching the loop between the tensioned rope and the top bar.

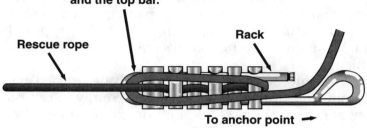

Rescue rope

Rack

To anchor point ➔

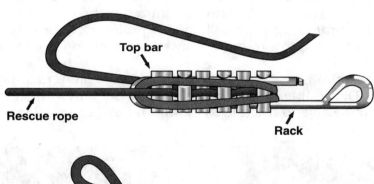

Top bar

Rescue rope

Rack

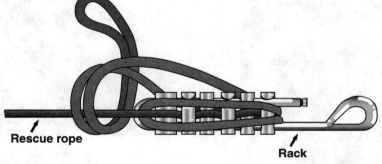

Rescue rope

Rack

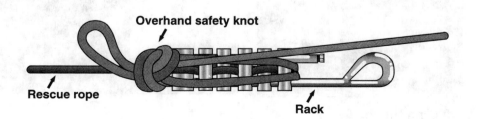

Overhand safety knot

Rescue rope

Rack

Making the Rescue

Now that you have arrested your descent and have communicated your intentions to the victim, it's time to effect the rescue. Position yourself around the bottom of the windowsill. You need to get the victim to sit on your lap. Some are willing participants and some aren't. The victim must sit on your lap, facing you and wrapping his arms around you. Make sure that the victim's position doesn't interfere with the descender.

You may encounter a victim who wants to be rescued but who is afraid to go out the window. If conditions permit, you can move into the window and have the person climb on from inside. Then, swing out and proceed to the ground. Be sure to size up any sharp edges. Once you cross the plane of the window, the rope may contact an edge. The chance of this is usually minimized if the rope from above is run over a projecting edge. While this projected edge may lessen contact with the window frame, it may also hinder you from reaching the window.

Once the victim is on board, tell him to remain calm, then proceed downward. Depending on the situation and the length of the rope, you may not be able to reach the ground. In most cases, all you really

Effecting the rescue.

need is to get several floors below the fire to reach safety. Interior companies should be in place to get you through a window or other portal.

This technique is for life-and-death situations only. The victim is secured only by his ability to hold on to the rescuer. When performed in training, the firefighter pretending to be the victim should be harnessed and secured to a belay system.

Harness Rescue

This is the preferred method of descending with someone, since both persons are secured to the system. This situation can be handled similarly to the no-harness rescue, except that it should go smoother than with a civilian. There may be situations in which a firefighter is injured or a civilian has been put into a harness for some reason and needs to be picked off. In this scenario, the victim is connected to the friction device of the rescuer. Instead of riding on the rescuer's lap, the victim is suspended below. The descending firefighter must bring a rescue sling and extra carabiners.

As before, stop just above the victim and communicate your intentions. Then, stop and tie off as quickly as possible in a position where you can reach the victim. Use a presewn rescue sling about four to five feet long. One end is connected with a locking carabiner to the friction

(Left) Tie off in a position where you can reach the victim. (Right) The rescue sling should be about four to five feet long.

(Left) After the sling has been connected, the victim should sit on the sill with both legs hanging out the window. (Right) Descend slowly when making a harness rescue.

device or to the locking carabiner between the device and the harness. It is important *not* to connect the rescue sling directly into your harness connection point. This may stress the harness in a manner for which it was not designed. Also, injuries could occur to the rescuer.

Next, reach over and connect the other end of the rescue sling to the victim's harness connection point with a locking carabiner. Tell the victim to get both legs out and sit on the sill. Assist the victim out of the window by pulling on the rescue sling and positioning it between your legs. Make sure it's clear of the window and upright between your legs. Untie or unlock your descender and slowly proceed with the descent.

EQUIPMENT PREFERENCES

Throughout this chapter, I've tried to cover the operating parameters of the figure 8 and the brake bar rack. My own preference is for the figure 8 for short-distance evolutions. There are many times when the brake bar rack is superior for developing friction; however, getting it over the operating edge presents a dangerous problem to the operator. Long before the brake bar rack became popular with the fire service, the rescue-grade figure 8 served quite well.

As an example of real-life rope evolutions, the Fire Department of New York uses a team approach to performing a rope rescue. In general, three members are required. One controls the descent with a fixed-brake arrangement, another is lowered to rescue the victim, and the third maintains communications between the other two. Their equipment consists of a life-saving rope, personal harnesses, and/or a life belt. The roof man ensures that the rescue equipment gets to a point above the victim. The rope used is 150 feet of 9/16-inch nylon laid with a minimum breaking strength of 9,000 lbs. Hooks are spliced into both ends. The rope is stored and transported in a backpack, which also contains an antichafing device. The personal harness can be used by the lowering man and the descending firefighter.

The system is created after assessing substantial objects to which the lowering man can secure himself, as well as the friction device. One hook of the life-saving rope is attached to the pompier hook of the harness that's donned by the lowering man. Additional slack is pulled to tie off to a substantial object, facilitating a secure point for the friction device and the firefighter operating it. The other end of the rope is attached to the descending firefighter. At this point, you have one firefighter attached to an anchor at one end and another attached to the member going over the edge. The descending firefighter sizes up the roof or window edge area and the victim's location. The third firefighter communicates the commands to the lowering man to ensure prompt and accurate positioning. The descending firefighter makes the grab with both hands. The victim sits on the lap of the firefighter, holding his upper torso. The topside firefighter then communicates further instructions to the lowering member. The descending pair may then be lowered to the ground or another safe area.

LEAPFROGGING

Leapfrogging is also known as retrievable rappelling. It is used when you have only one rope and long distances to cover from either a high- or a low-angle position. The basic premise of this technique is that, in an emergency, you can rappel to an interim location, detach the rope, fix the end to a new anchor point, drop the other end, and continue the descent. It is considered a single-rope technique because, if the anchor fails or the rope breaks, there is no backup.

To start, run the middle section of the rope around a bombproof anchor point at the top of the rappel, then tie the two ends together. Make sure the anchor doesn't bind the rope in any way, since ultimately the rope will be pulled around it from below. Out in the wilderness, a tree trunk will work. Toss the knotted end of the doubled rope over the precipice, and rappel down to a securable location, preferably above the knot. In most cases, you must secure yourself to a new anchor before coming off the lifeline. Once disconnected, untie the knot and pull the loose end of the rope through the anchor above, then retoss that end from your current location and continue the descent. You can repeat the process as many times as necessary to reach the bottom. Inspect the rope between stages, since pulling it around the anchor point from an inferior position may chafe it. Generally speaking, any webbing and hardware you use along the way to create new anchor points will be lost, since it'll be impossible to get back up to retrieve them.

KNOT BYPASS

Some operations require tying two lifelines together, resulting in a knot that your descender cannot pass. The technique for getting around this knot, or any natural knot that develops, involves one long and two short prusik loops, plus extra carabiners. Some experts also use ascenders and accessory cord.

We'll use a scenario in which no prusik hitch safety is in place. First, stop just above the knot and tie off the descender. Then, place two prusiks to the rescue rope, one above the descender and one below the knot. Attach these to a locking carabiner that's connected to your harness. Untie the double wrap on the descender and transfer the weight to the top prusik. Disconnect the descender and reconnect it below the knot. Rewrap the descender twice as required and check it for security. To remove the tension from the top prusik, place a foot-loop prusik on the rescue rope above the knot. While standing in the foot loop, verifying that your descender is securely wrapped, disconnect the initial top prusik from your harness. Step out of the foot loop, transfer your weight to the rescue rope, and unwrap it. Remove either two or all three prusiks. You need to remove at least two; however, you can leave the lower one in place to use as a self-belay.

PERSONAL ESCAPE SYSTEMS

Personal escape systems were formally recognized under the 1995 edition of NFPA 1983. The question as to which size and capacity of

A personal escape rope kit is indispensable equipment for the rescue firefighter.

rope to carry had been talked about for years in fire stations. Although many firefighters had been carrying various types of rope and webbing in their turnout coats anyway, the rope standard finally legitimized doing so. Manufacturers have since produced different versions of escape ropes. These aren't just prusik cords rolled up and stored in someone's pocket. Many of these products are 5/16-inch (8 mm) rope with breaking strengths of 3,000 lbs. They are one-person components with a maximum working load of 300 lbs.

The personal escape rope is one component of an escape system. Several manufacturers produce complete escape packages for firefighters to carry during the course of their duties. These run from the very basic to the more complex in terms of components and the techniques that they require.

As always, personal escape systems require special training, and you must follow the manufacturer's recommendations.

STUDY QUESTIONS

1. SRTs originated with _____ because of the long drops they encounter and their need for simple systems.
2. Name the seven components of the acronym RESCUES.
3. When rappelling, most rescuers use their dominant hand for _____.
4. When rolling over a parapet during a tie-low evolution, where must your guide hand be in relation to the rope?
5. When rappelling down to a window during a no-harness rescue, keep your brake hand _____ from the victim.

Chapter 12

Mechanical Advantage Systems

In rope rescue systems, mechanical advantage is needed anytime you have to raise a victim or firefighter to a higher location. Doing so requires a solid understanding of mechanical advantage theory, as well as additional equipment and skills. In this chapter, we will cover the necessary fundamentals of hauling, then illustrate some of the common systems in use.

Any mechanical advantage gained in a system is always less than its theoretical potential. Whenever mechanical advantage is calculated, it's done on the premise that all of the components in the system are frictionless, resulting in TMA, or theoretical mechanical advantage. Practical mechanical advantage represents the actual advantage gained over a number of physical variables, including the friction of the pulley and rope, for example.

Suppose that a team responds to an incident in which a maintenance worker is down and injured at the bottom of a utility shaft. The only opening is at the roof, and the worker is down approximately 100 feet. After firefighters descend to the victim, stabilize him, and package him for transport, they decide to haul him out. There is an adequate steel superstructure above the opening, and they attach a single pulley to it. They then attach a static rescue rope to the victim's harness and reeve it through the topside pulley. The victim weighs approximately 300 lbs., including an oxygen kit, plus splinting materials. The hauling team, which consists of four firefighters, struggles to pull him up. Not only do they find it difficult to bear the weight, they also have no mechanism to hold him when they need to reset the system. Mechanical advantage is quite necessary in this situation due to the distance of travel and the weight of the load. The single pulley only provides a means to change the direction of the hauling rope. When the haul team pulls, their effort equals the weight of the victim, plus a little more to overcome whatever friction is in the system. In theoretical terms, this system offers no advantage whatsoever. The ratio of the useful output of the machine to the force applied is less than 1:1.

This scenario illustrates our basic need for mechanical advantage systems—i.e., machines to perform work. The work output of your system should be much greater than the effort put in. For hauling, mechanical advantage requires the use of pulleys, which can be placed in various configurations to achieve different ratios. Pulleys that are permanently anchored or that change direction *do not* provide mechanical advantage. These are known as directional, fixed, anchored, or stationary pulleys. Mechanical advantage is provided only by traveling pulleys—i.e., those that move within the system.

In each of the examples below, the mechanical advantage is 1:1. The placement of the pulleys does nothing more than change the direction of the rope. A 300-lb. load would need 300-plus lbs. of force. The last arrangement may require more effort because of additional friction created by the Z drag.

1:1 Mechanical Advantage Systems

In the next arrangement (page 225, top), the pulley is attached to the load instead of the overhead anchor point. The placement of the pulley on the load allows for the weight to be distributed between the two legs of the system.

A 2:1 mechanical advantage system is the most basic of the systems wherein practical mechanical advantage is actually gained. The rope is attached to an anchor and reeved through a pulley on the load, then brought back up through a pulley at the superior position. The 300-lb. load is being divided in half, with 150 lbs. on each leg of the system.

Load Divided Between the Legs of a System

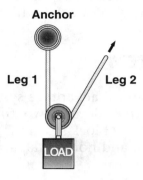

2:1 Mechanical Advantage

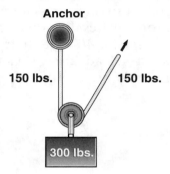

One of the primary attributes of the pulley is its ability to equalize the load between the legs of a system. The effort put in needs to be a little more than the theoretical amount to overcome both friction and inertia.

This setup is typical of a simple mechanical advantage system, which by definition consists of one rope and one or more pulleys. Such a system provides all of its own mechanical advantage. By adding pulleys and/or reeving more sheaves, you can increase the benefit. In the example given, the advantage can be increased from 2:1 to either 3:1 or 4:1 by adding another pulley and/or reeving another sheeve. An easy way to calculate the mechanical advantage of a simple system is to count the number of ropes that are supporting the load, in this case two.

A compound system is comprised of two or more simple systems working together to multiply the overall capability. A 2:1 simple system added within another 2:1 simple system yields a compound system of 4:1 advantage. Whenever you add or stack simple systems, multiply their inherent ratios to calculate the overall advantage.

THE BLOCK SYSTEM

For centuries, mechanical advantage systems have commonly been of block and tackle, with pulleys of two, three, and four sheaves being the norm. Even today, the block system is highly versatile, and it can be used in both vertical and horizontal arrangements.

The Vertical Arrangement

These types of systems are popular in confined space operations where someone is lowered into a vertical space. Once the firefighter is secure below, either a mechanical ascender or the cam device of a combination pulley is installed so that the system will be set for a haul evolution. The length of rope relates directly to the mechanical advantage of the system. For example, a 40-foot shaft requires 160 feet of rope (plus tails) to produce a 4:1 advantage. Generally speaking, the 4:1 ratio is the most commonly used setup.

3:1 Block System

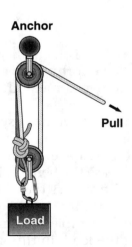

The working end of the rope gets secured to the carabiner that's holding the pulley to the load.

4:1 Block System

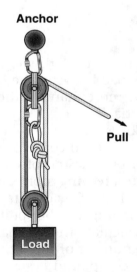

The working end of the rope gets secured to a carabiner affixed to the anchored portion of the system.

The advantages of the vertical arrangement are:

1. It requires limited equipment.
2. It requires limited operators.
3. It can develop high mechanical advantage ratios and capabilities.

The disadvantages of the vertical arrangement are:

1. It requires long lengths of rope.
2. High mechanical advantages can overstress the components in the system, endangering the load.
3. In a rescue environment, it must be backed up with a belay line and/or a redundant system to an independent anchor point.

The vertical arrangement has limited capabilities in situations where you encounter great depths. It remains, however, a popular and useful system and is just one of the many tools available to you.

The Horizontal Arrangement

The horizontal arrangement is used extensively in piggyback evolutions (see below). When a victim or a firefighter is hanging from a rescue rope, a horizontal block system can be installed to effect the vertical raise. A pulley system is fixed to a firm anchor, then run horizontally to attach to the standing rope. The mechanical advantage of this type of rigging is calculated in the customary way, by counting the number of ropes passing through the movable pulley.

THE Z SYSTEM

The Z system is a highly versatile and relatively easy system to set up. It requires one main haul rope and two pulleys, plus two rope grabs attached to the haul rope. This arrangement typically uses two single-sheave pulleys for a 3:1, or two double-sheave pulleys for a 5:1, mechanical advantage. Depending on available personnel, opting for a 5:1 ratio isn't a bad idea. Use the double-sheave pulleys right away, even if you're starting out with the 3:1. In the event that you need the additional mechanical advantage, you can wrap the other two sheaves and increase the system to a 5:1 ratio. As the rope grab travels upward toward the anchor pulley, the system needs to be tied off and the rope grab reset, a disadvantage in terms of time. The benefit of this type of rigging over a block system is the amount of rope required to gain the

3:1 Z System

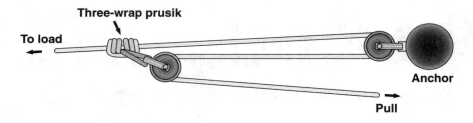

5:1 Z System

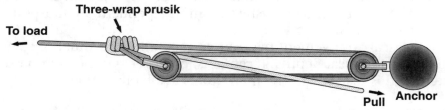

**This arrangement differs in appearance from the typical
Z system; however, it is still classified as one.**

mechanical advantage. In a 100-foot rescue scenario, a Z system might
require less than 200 feet of rope, depending on the desired length of
the legs in the system and how many resets are permissible.

The advantages of a Z system are:

1. It requires limited equipment.
2. It requires limited personnel.
3. It is a simple mechanical system that can be reset.
4. It can be expanded into a compound system.

The disadvantages of a Z system are:

1. Small operating areas may require many resets to achieve the
    desired hauling results. This may be offset by placing a change-
    of-direction pulley between the system and the haul team.
2. High mechanical advantage capabilities can overstress the sys-
    tem components, endangering the load.
3. In a rescue situation, it must be backed up with a belay line
    and/or a redundant system to an independent anchor point.

THE PIGGYBACK SYSTEM

The piggyback system is an arrangement wherein a mechanical advantage system is connected to a main haul line. A typical example would be to piggyback a rescue rope on which a firefighter descends to reach a victim, offering mechanical advantage to haul them both out again. You can connect a simple system or stack systems to create a multisystem advantage. In a 2:1 piggyback, a pulley is attached to the haul line as close as possible to the load or, more likely, near the edge of the operating area. The 2:1 rope system is then fixed to an anchor point. For every two feet you pull, you'll move the main line one foot, but the force exerted will be half the weight. Other advantageous ratios can be rigged, as shown in the diagrams, the classic one probably being the 4:1 compound system piggyback. This arrangement is put together by stacking one 2:1 simple system onto another 2:1 simple system. For better allocation of resources, the piggyback can be rigged with one rope rather than two. This results in the same attributes of other 4:1 compound systems while using fewer pulley sheaves and less rope.

THE ODD AND EVEN PRINCIPLE

In determining ratios of mechanical advantage, your system will be either odd- or even-numbered. To figure out which you have, remember that if the working end of the rescue rope is secured to the anchor or anchor pulley, then the system is even. If the working end of the rescue rope is secured to the load or load pulley, then the system is odd.

CHANGE-OF-DIRECTION PULLEYS

There have been many references to change-of-direction pulleys in this chapter. The primary emphasis has been that they don't create or contribute mechanical advantage to a system. The common misconception that arises is that "if it's not creating mechanical advantage, then it's not an important component." Nothing could be further from the truth. The change-of-direction pulley allows you to run your line from an area of limited space to an area where a haul team can operate more efficiently. Depending on the angle of the rope going through the pulley, the load on the anchor point can be increased up

1:1 Piggyback System

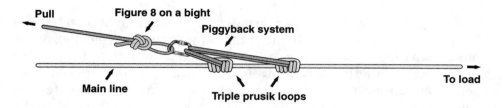

2:1 Piggyback System

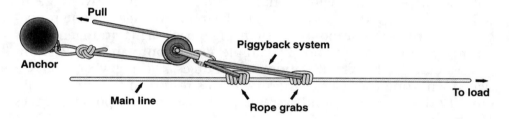

3:1 Piggyback System

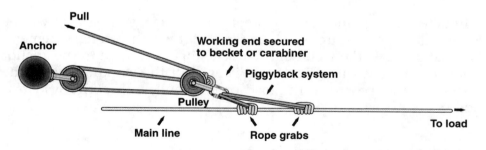

4:1 Piggyback System

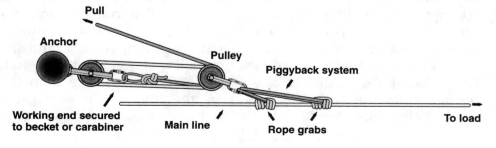

4:1 Compound Piggyback System

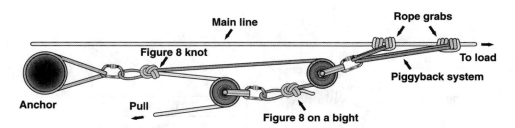

Main line

Rope grabs

Figure 8 knot

To load

Piggyback system

Anchor

Pull

Figure 8 on a bight

5:1 Piggyback System

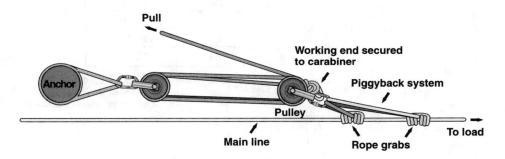

Pull

Working end secured
to carabiner

Piggyback system

Anchor

Pulley

Main line

To load

Rope grabs

to two times the weight of whatever you're hauling. This is because a pulley has to withstand the weight of the load on one side and the effort being expended on the other. Anchor points must be chosen carefully and backed up accordingly.

OVERSTRESSING THE SYSTEM

Of the many problems that can occur in a mechanical advantage system, one of the most apparent is a consequence of high mechanical ratios. The higher the advantage, the easier it is to pull up the load. Given the limitations of equipment, there has to be a balance between the system's capacities and the effort expended by the haul team. In general practice, the fewer haul members available, the higher the ratio required. Suppose you'd set up a 5:1 Z system because staffing had been limited to two members. Halfway into the operation, three additional members appear on the scene, ready to help. If you figure that the average brute can exert between 100 and 200 lbs. of pulling force, five members working as a team can reach or exceed 1,000 lbs.

Change-of-Direction Forces

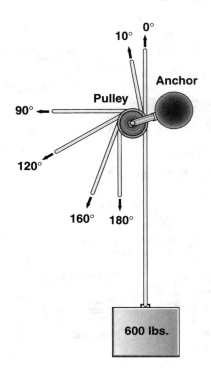

600 lb. Load	
Angle	**Load imposed**
0°	0 lbs.
10°	204 lbs.
90°	600 lbs.
120°	846 lbs.
160°	1194 lbs.
180°	1200 lbs.

Load imposed = Stress put on pulley
and its anchor point.
Select components carefully!

This is a dangerous situation in the making. A wide variety of smaller problems can occur—a stokes getting caught on an edge or a rope tangling around an object, for example. Once these firefighters meet some resistance, the natural tendency will be to dig in and pull harder. This can cause failure of the system and loss of the victim or rescuing firefighter. Use great care and communicate freely, particularly when hauling through narrow openings. Overstressing can also mean situations where human limbs get wedged between substantial objects, possibly getting mangled and torn by the application of too much mechanical advantage.

You'll see compound systems of 10:1 advantage or even more, depending on how the components have been stacked. To ensure safety, some teams won't go beyond a 5:1 ratio unless doing so is absolutely justified. A 5:1 system with two haul members pulling continuously is usually sufficient for all two-person rescue loads.

STUDY QUESTIONS

1. What is the most popular and advantageous mechanism for confined space rescue?
2. Name the arrangement in which a mechanical advantage system is connected to an existing main haul line.
3. If the working end of the rescue rope is secured to the anchor or anchor pulley, then the system is _____ .
4. Any change-of-direction pulley must be able to withstand two opposing forces. What are they?
5. To ensure safety, some teams won't rig systems beyond a mechanical advantage ratio of _____ unless doing so is absolutely justified.

Chapter Thirteen

Transportation Systems

ooking at the life safety system structure reveals three major subsystems: the anchor system, the access system, and the transportation system. This chapter will cover the principles of transportation and its two subcategories, lowering systems and hauling systems using mechanical advantage.

LOWERING SYSTEMS

Because they take advantage of gravity, lowering systems pose limited demands in terms of equipment, requiring only an anchor system, a descent device and rescue rope, a redundant backup system, and some type of victim transportation device, such as a stokes. In the chapter on descending, we discussed rappelling to a victim and "making a grab." Even though this was discussed under the auspices of the access system (descending), most victim handling is done within the transportation subsystem. This aspect of rescue provides a mechanism to move a victim from one location to another, as well as a patient handling device.

Lowering and descent evolutions are essentially quite similar, depending on their application. Both rely on a brake to control the amount of friction required to manage the load. In the fire service, evolutions that rely on traveling brakes are usually termed descending operations, whereas those that rely on fixed brakes are usually termed lowering operations. This latter group will be the focus of this chapter.

A lowering system is most often used for the evacuation of an injured person during nonfireground technical rescues. This type of evolution requires a main lowering line and an independent, redundant backup system. The personal descent is primarily designed to allow a firefighter to reach an inferior position no matter what the ultimate course of the operation. The most significant aspect that distinguishes these two techniques is that the personal descent can be set up as an SRT evolution, while lowering automatically entails a DRT evolution.

The accompanying diagram shows a basic arrangement using a figure 8 ring. Since this is a double-rope technique, a separate and independently anchored backup system must be set up adjacent to the main one—it has been eliminated from the diagram for clarity. This example depicts one of the many ways you can construct this evolution in the field. The necessary components are the anchor points, the anchor system, the friction device, static rescue rope, edge protection, and, as an operational component, rope management. Whenever you construct a lowering or hauling system, two ropes are usually required at minimum. Such systems are either redundant or parallel in nature. Redundant means having one primary line that takes the brunt of the

A Basic Lowering System

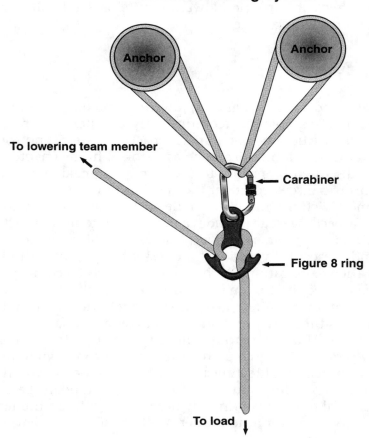

To lowering team member

Carabiner

Figure 8 ring

To load

load and a backup that serves only in the event that the primary fails. A parallel system divides the load between two identical evolutions that work in unison. In theory, a 300-lb. load would impose about 150 lbs. on each system. In actual practice, coordinating two teams to lower or haul at the same rate in a parallel system sometimes doesn't work out. One side of the double system always seems to be doing more of the work; hence, you wind up with a redundant system anyway. This, in part, is a training issue. In Chapter Three, we discussed rescue loads in terms of their imposition on a system. Limiting a load to two persons or 600 lbs. is prudent. Using 1/2-inch static rescue rope as the primary and backing it up with a secondary system gives you a tremendous margin of safety in the event a problem occurs.

The brake bar rack arrangement is similar to the figure 8 ring. It is inserted as a fixed brake device. It must be threaded correctly and all the bars set for maximum friction during the initial stages of the evolution.

A majority of teams carry and deploy their ropes from nylon rope bags. Depending on the evolution at hand, as well as the size and complexity of the operating area, you may want to pull the rope from the bag and prestage it. This can be helpful for long drops and whenever additional personnel are available. Postinspection and proper repacking become doubly critical whenever ancillary personnel are involved.

In the discipline of rope rescue, operations are classified as either high angle and low angle. High-angle evolutions denote situations in which the firefighter or victim is supported solely by the rope system. Low-angle evolutions refer to situations in which the individual still has contact with the ground and only partially depends on the line for support. The vast majority of rope rescue operations take place within the high-angle environment; however, low-angle evacuations from ditches, drainage systems, icy slopes, and the like are common. In a typical low-angle scenario, a victim is secured to a stokes basket, which four to six firefighters must haul with their own muscle power. The slope is steep enough that hanging on to the stokes while maintaining balance becomes impossible; consequently, a low-angle evolution is arranged. Slope evacuations can take place both up and down elevated topography.

The Downward Slope

Moving from a higher to a lower position tends to be easier and should be the rescue team's first choice. Considerations include terrain features and the availability of anchor points. Also, the brakeman at the anchored friction device must be able to see the entire progression. If not, radio communications must be employed, increasing the potential risks. Vegetation and rocks that could endanger the rescue line are

Downward-Slope Evacuation With Rack

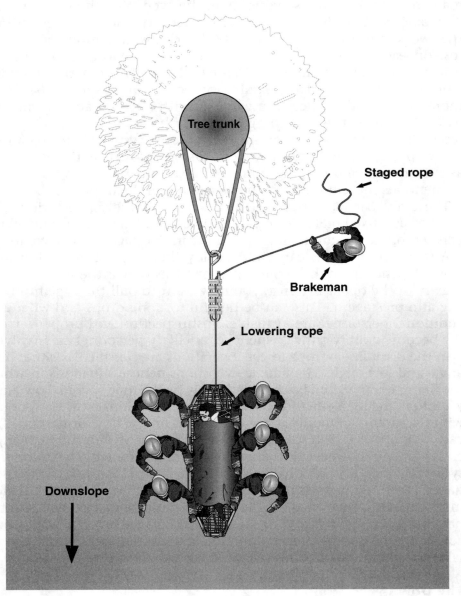

always considerations. You must locate anchor points along the way at a rapid pace, particularly in a downward evacuation.

The first slope evacuation technique I learned was taught to me by U.S. Air Force pararescue personnel. It was designed for a rapid tactical evacuation. In this method, a team carries a victim in a stokes while

Tree Brake Wrap

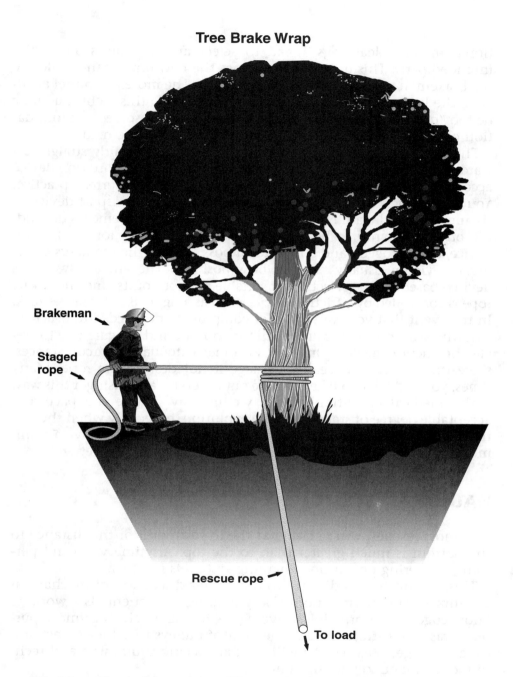

Brakeman

Staged rope

Rescue rope → To load

assisted by a lowering system. The system consists of a static rescue rope attached to the head of the stokes and to a solid anchor point. The friction device is a munther hitch around a carabiner attached to a tree by some webbing. While the team descends the slope, an additional member leapfrogs ahead to select anchor points along the intended path. This individual becomes the new brakeman, while the old brakeman becomes the new leapfrogger. The more personnel available, the less strain on these two members. While this technique may be too "quick and dirty" for nonmilitary rescues, it serves as a foundation on which improvements and adjustments may be made.

The tactical considerations of such a technique are fairly straightforward. As always, you must secure the victim in the transportation device according to the manufacturer's recommendations or current practice. You must attach the rescue rope to the head of the transport device. A six-member haul team is typical and may require replacements depending on the distance to be covered. Select a brakeman for the friction device, a leapfrogger, and a haul team boss. One of the members carrying the victim usually fills the boss position. The anchor system is devised based on the selection of suitable anchor points, and the rescue rope is controlled by a friction device such as a figure 8 or brake bar rack. In the event that you were short of equipment and needed to evacuate a downward slope, you could wrap the rope around a secure tree to create the friction device. Only do so under extenuating circumstances, since the abrasion can easily damage if not destroy the rope. Many times, you will have to take the rope out of service after using it this way.

The medical condition of the victim may dictate the pace and acceptable degree of stability of the evolution, especially when the victim requires constant medical attention. All teams must have a minimum of two basic EMTs; however, paramedics would be better.

HAULING SYSTEMS

In most rescues, you're either at the lowest point or the distance to the bottom is much greater than to the top. Anytime you can't perform a lowering operation, a hauling system is in order.

This section relates directly to our earlier discussion of mechanical advantage. The construction of any hauling system entails a working knowledge of the principles involved. Chapter Twelve's primary purpose was to illustrate the arrangement of pulleys to achieve mechanical advantage. This section will look at specific equipment and techniques germane to hauling systems.

Rope Grabs

Rope grabs are commonly used in hauling systems either as a means of attaching pulleys or as ratchets for hauling. Both soft ascenders such as prusiks and hard ascenders such as mechanical ascenders are used. The advantages and disadvantages of both take on additional significance in hauling systems. Human preferences run in cycles, and the issues surrounding hard and soft ascenders are no exception. Although the hard ascender's popularity has grown immensely in recent years, the prusik loop has come back into favor among many teams. Some have totally replaced the hard ascender with prusiks, while others work with them in a hybrid arrangement. Prusiks are often used as cams in the haul or ratchet function. A haul cam grips the rope and is connected to a traveling pulley. The ratchet cam is usually positioned back by the anchor, holding the system from sliding when the haul cam needs to be reset.

The Load-Releasing Hitch

The load-releasing hitch has gained popularity because it offers flexibility in loosening up a tensioned system. Placing it in a hauling system allows for quick conversion to a lowering evolution. This hitch can be made from a variety of accessory cords, webbing, and rescue rope.

Block Systems

The block system can be used in both a vertical and a horizontal arrangement. In a vertical 3:1 system, a hard mechanical ascender

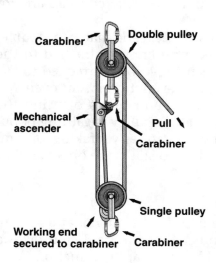

3:1 Block System

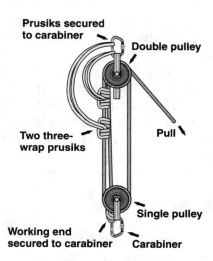

3:1 Block System With Prusiks

4:1 Block System With Combination Pulley

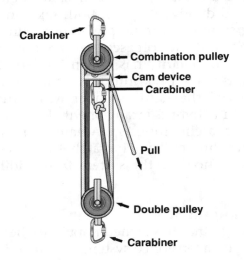

attached to the rope acts as a brake so that you don't lose whatever lift you may have achieved. This device needs to be minded at times, since it may bind up against the pulley. Any personnel near the opening need to be properly tethered to the anchor in the event of a fall. A 3:1 block with a combination pulley and cam device eliminates the need both for the ascender and for someone minding it. A 3:1 block with prusiks is used by some teams. Generally it must be minded, since the grip is bidirectional, and the firefighter must be tethered to an anchor when operating near an opening.

A 4:1 arrangement is very similar to a 3:1, except that two double-sheave pulleys are required, and the securing knot is attached to the top pulley. Remember that when the securing knot is attached to the load, it's an odd system; if it's attached to the anchor, it's an even system. This setup can be arranged like the 3:1 in terms of using the ascender or prusiks; however, the combination pulley is gaining popularity. In the event that you don't incorporate any braking or cam device near the pulley, you can still arrange a distal brake on an anchor point in the operating area.

Z Systems

In a 3:1 Z system, hard ascenders are used in the ratchet and at haul positions on the hauling rope. Remember that the haul cam takes a bite and grips the rope while it's pulled back toward the anchor. Once

3:1 Z System With Gibbs Ascenders

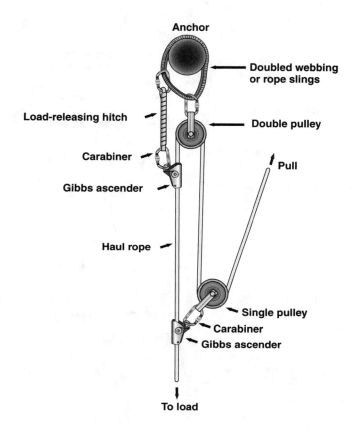

the two cams reach each other, the ratchet cam holds the load while the haul cam is released and set forward to take another bite. Some teams advocate using an additional cam to act as a safety. Many omit the safety cam, primarily because they back up the haul system with an independently anchored safety system. In such a case, the ratchet cam works as a safety cam when the hauling cam is operating.

In a 3:1 Z system with prusiks, tandem prusik loops are used in the ratchet and haul positions. Rescue prusik cord and 8 mm accessory cord are popular on 1/2-inch static rescue rope. It's best to use a short and a long loop together so both prusiks can grip at the same time. This arrangement may take a little longer to operate and reset than others.

Hybrid 3:1 Z System

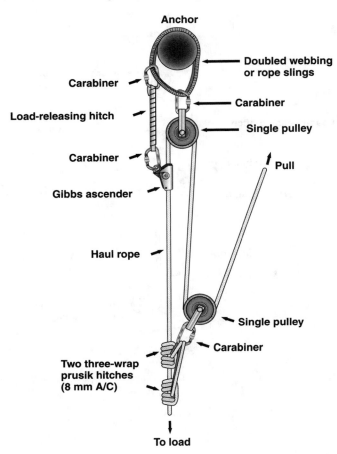

In a 3:1 hybrid arrangement, prusik loops are used at the haul cam position, and a hard ascender is used at the ratchet. Those who use this approach are uncomfortable with the resultant manipulation of the haul rope when it is gripped by a hard ascender.

The 5:1 Z system is similar to the 3:1, except that two double-sheave pulleys are required to produce the increased mechanical advantage. This system can be arranged like the 3:1 in terms of using the ascenders, prusiks, or both.

Piggyback Systems

The premise of a piggyback is that a static rescue rope is attached to the load you wish to haul. The system is then attached to that rope by a rope grab. This results in operating characteristics different from either the

2:1 Piggyback With Gibbs

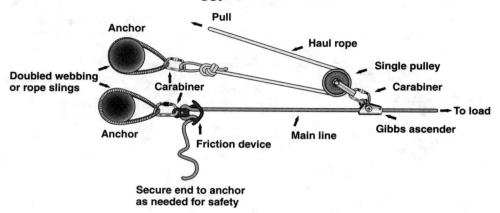

3:1 Piggyback With Prusiks

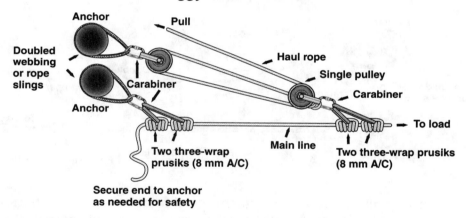

4:1 Piggyback With Prusiks

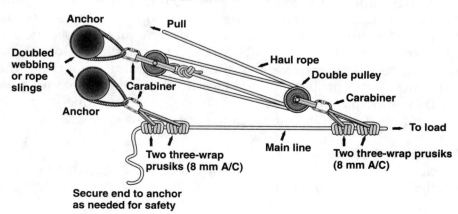

4:1 Compound Piggyback System

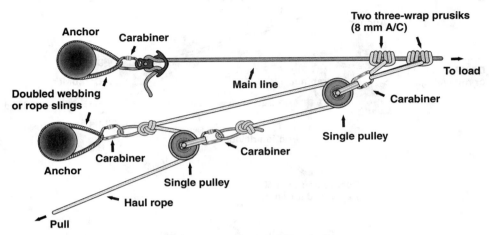

5:1 Piggyback With Prusiks

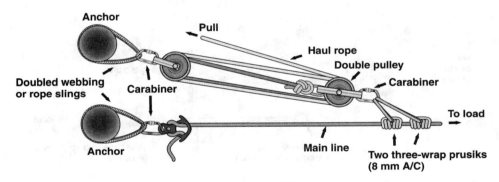

block or Z systems. Those arrangements need to be backed up with a second rope so as to comply with the DRT principle. In a piggyback setup, the hauling system is attached to the main line, forming a cohesive unit.

The main line still serves an important function when the system must be reset. As the load is being hauled, rope is being pulled back through a friction or brake device in the event a problem occurs. Some teams consider this to be a DRT arrangement. Still, the load is only being supported by one rope past the rope grab attachment, and there is no backup if a problem occurs beyond this point. To comply with the DRT principle, you must attach a second rescue rope to the load and connect it to an independent anchor system as a backup.

The Upward Slope

Evacuating a slope in an upward direction requires either personnel to effect a direct haul or a mechanical advantage system. Tactically,

1:1 System With People Power

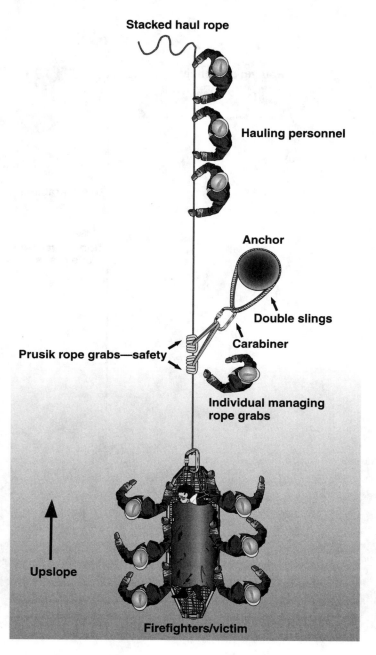

Stacked haul rope

Hauling personnel

Anchor

Double slings

Prusik rope grabs—safety

Carabiner

Individual managing rope grabs

Upslope

Firefighters/victim

such an operation is similar to lowering; however, the friction devices are either prusiks or hard ascenders, and leapfroggers aren't necessary.

1:1 Change-of-Direction System With People Power

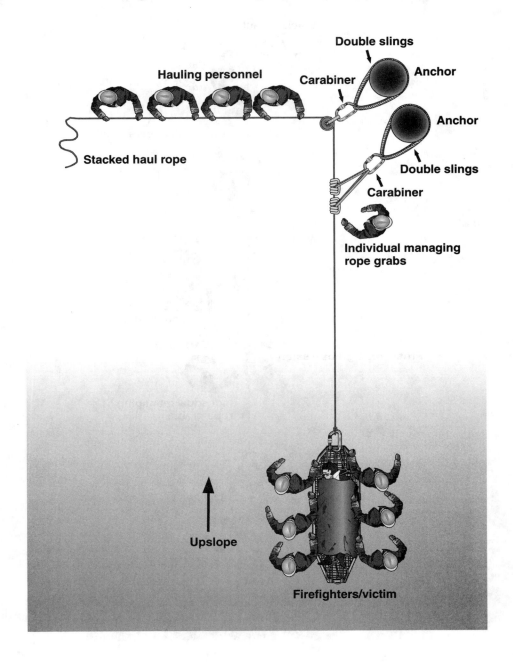

2:1 System With People Power

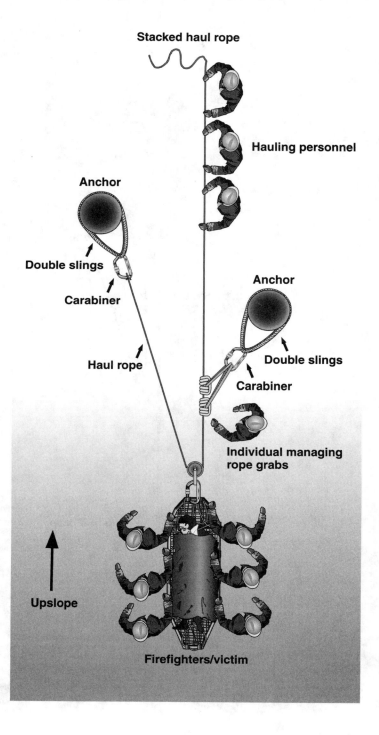

Stacked haul rope

Hauling personnel

Anchor

Double slings

Carabiner

Haul rope

Anchor

Double slings

Carabiner

Individual managing rope grabs

Upslope

Firefighters/victim

Topographic Hauling Evolution

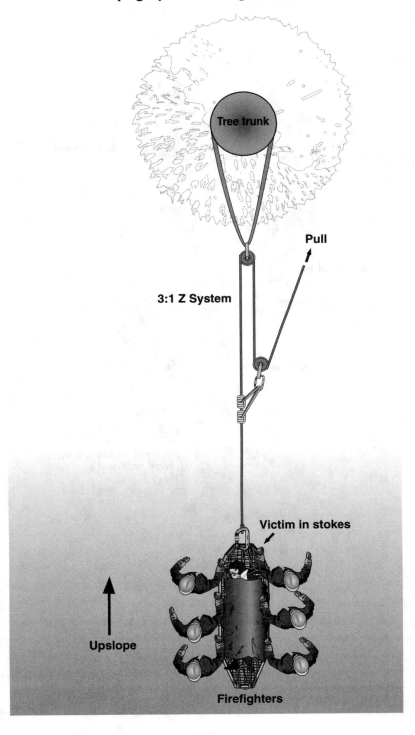

THE CONVERSION EVOLUTION

There are times when you need to go from a lowering/descending evolution to a hauling evolution, or vice versa. Suppose that a firefighter descends to an inferior position to effect a rescue. Once there, the firefighter and the victim need to be hauled upward to safety. They may be tethered to an anchor point, rest safely on the bottom, or be dependent on the lowering/descent system. In any case, the topside team members can convert the lowering system to a hauling system with a few simple maneuvers. Rearranging it is somewhat easier, of course, if the load can temporarily be taken off the rescue rope.

To convert a lowering/descent evolution to a haul system, tie off the friction device with a securing knot, and tie a doubled-up single fisherman knot as shown. Then, attach a rope grab to the main line with a load-releasing hitch attached to an anchor point, if one isn't in place already. Next, tie a safety figure 8 in the section of rope between the friction device and anchor point, and attach it to the anchor point. Remove the friction device from the lowering/descent system. Reeve the required pulleys for the desired mechanical advantage, and secure them until the conversion is complete. Release the load with the load-releasing hitch. This will impose the load onto the newly created mechanical advantage system. Finally, reset the rope grab to use as a system cam device.

**Converting From a Lowering/Descent System
to a Hauling System (Before)**

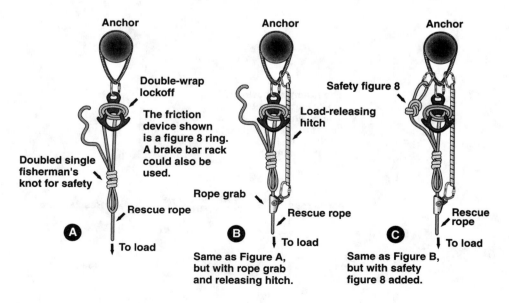

Converting From a Lowering/Descent System to a Hauling System (After)

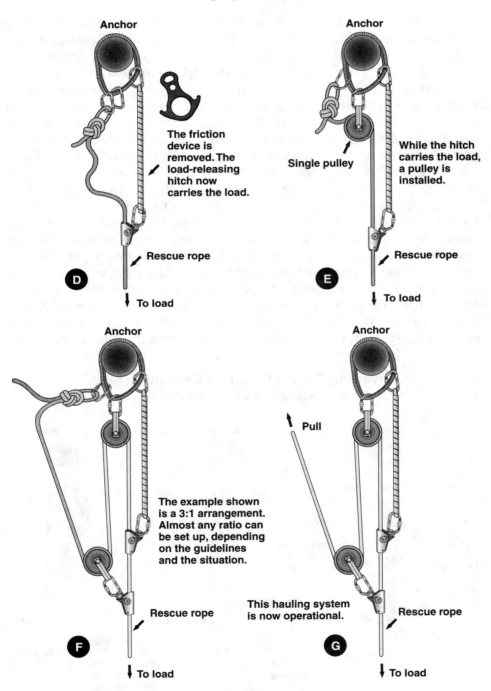

Anchor

The friction device is removed. The load-releasing hitch now carries the load.

Rescue rope

D

To load

Anchor

Single pulley

While the hitch carries the load, a pulley is installed.

Rescue rope

E

To load

Anchor

The example shown is a 3:1 arrangement. Almost any ratio can be set up, depending on the guidelines and the situation.

Rescue rope

F

To load

Anchor

Pull

This hauling system is now operational.

Rescue rope

G

To load

Converting From a Hauling System to a Lowering System (Before)

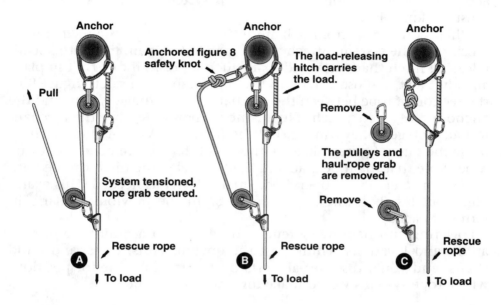

Anchor

Pull

Anchored figure 8 safety knot

Anchor

The load-releasing hitch carries the load.

Anchor

Remove

The pulleys and haul-rope grab are removed.

System tensioned, rope grab secured.

Remove

Rescue rope

Rescue rope

Rescue rope

A ↓ **To load**

B ↓ **To load**

C ↓ **To load**

Converting From a Hauling System to a Lowering System (After)

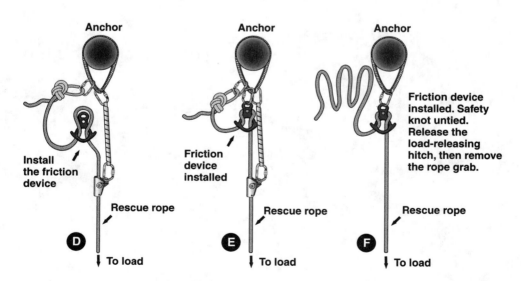

Anchor

Anchor

Anchor

Friction device installed. Safety knot untied. Release the load-releasing hitch, then remove the rope grab.

Install the friction device

Friction device installed

Rescue rope

Rescue rope

Rescue rope

D ↓ **To load**

E ↓ **To load**

F ↓ **To load**

So as not to confuse the sequence, this description doesn't take into account the conversion of the backup system. Remember that this must be done also.

When converting from a hauling to a lowering system, maintain tension on the system. Attach a rope grab to the main line with a load-releasing hitch that's attached to an anchor point, if one isn't in place already. Next, impose a load on the rope grab. Tie a safety figure 8 in the section of rope between the mechanical advantage system and the anchor point, then attach it to an anchor point. Remove the mechanical advantage pulleys and install the friction device of choice. Then, untie the safety figure 8. Release the load hitch and remove the rescue rope. Note that the rope grab and the load-releasing hitch can be reinstalled for safety if it's supervised. Finally, lower as needed. This example stipulates the same requirements as in the previous scenario in terms of backups.

Lowering and hauling systems can be used in a multitude of situations and locations. Think past the obvious cliffs, roofs, and blind shafts, and you'll discover all sorts of interior and exterior positions where rope systems would make for smoother operations.

STUDY QUESTIONS

1. In the fire service, evolutions that rely on traveling brakes are usually termed _____ operations, whereas those that rely on fixed brakes are usually termed _____ operations.

2. What sort of system divides the load between two identical evolutions that (theoretically) work in unison?

3. What is the name for a mechanism in which the rescuer or victim is supported solely by the rope system?

4. A hybrid arrangement of rope grabs incorporates both _____ components.

5. In a 3:1 block system, is the securing knot attached to the top or the bottom pulley?

Chapter Fourteen

Personal Ascending

Personal ascending is a technique that allows a firefighter to move higher on a vertical rope. It is generally used for short distances in the fire service and may take some time to master. It requires little equipment; however, it is physically demanding and requires dedicated coordination. The limit for many may be as little as 150 feet.

Ascending in the rescue field is generally used for reaching other ropes higher in the system or for self-rescue. This is truly an individual activity rather than a team event. In spelunking or mountaineering, it may be the safest or only route available to the climber. Cavers have been known to ascend well over 1,000 feet just to reach daylight again, and this technique is considered vital to their overall methods and skills.

While there are many individual techniques, certain principles are most often incorporated. Ascending is primarily a single-rope technique, and a Class II harness is the minimum required. Use a three-cam system. The cams can be either hard or soft. An important human factor is to push with your leg muscles for upward mobility rather than using just your arms.

A cam is basically a rope grab that gives you many choices. In mountaineering and caving, the handled ascender is popular because it is easy to use. In the fire service, these devices are relatively unknown. The Gibbs type of ascender is a popular all-around rope grab. Tying a secure loop in 8 mm accessory cord and then tying that into a prusik knot on the rescue rope is considered a suitable soft ascender. After working with prusiks, you will realize why it's a limited-distance technique. Moving the loaded prusiks takes great effort; however, it's safe, it requires few materials, and it will work in a pinch.

To start, don at least a Class II harness. A Class III with the chest

Basic Ascending Setup

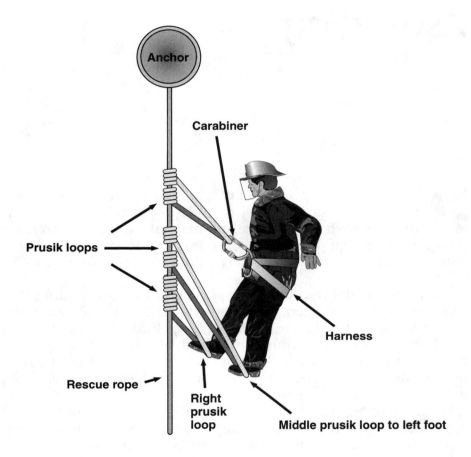

component would be better. Attach a prusik knot to the rescue rope, then run the distal end to your harness using a carabiner. Push the prusik above your shoulders. Tie two more prusiks to the rescue rope. You can attach either rope, webbing, or extended accessory cord to act as foot loops. Estimate the length needed for the legs. The basic challenge is to coordinate the movement of the three prusiks. The top one is for overall safety, the second is for the left leg, and the third is for the right leg.

Ascending Sequence

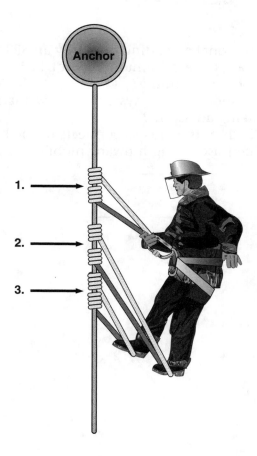

1. ⟶

2. ⟶

3. ⟶

Climbing requires you to move the leg cams upward alternately so that you can get a step effect. As you stand on one leg, bend the other, keeping it in contact with the foot loop as you move the prusik upward. Once you reach a certain point, load that leg and stand fully on the foot loop, then advance the prusik for the other leg. Avoid pulling yourself upward by your arms. Coordination is essential. The individual climber must decide the actual sequence of where to place and when to move the safety cam. Practice on short ascents to hone your technique.

STUDY QUESTIONS

1. Is personal ascending primarily an SRT or a DRT?
2. What is the minimum class of harness required for personal ascending?
3. You should use a system of how many cams when making an ascent?
4. Which of the prusiks connects to the harness?
5. When ascending, upward mobility is gained mainly by _____.

Chapter Fifteen

Traversing

Generally speaking, traversing entails moving horizontally from one elevated point to another—for example, from one building to the next. The evolutions for doing this are called by many names, the most common being tyroleans, telphers, and high lines. Once such a system is in place, it's safer and quicker to move personnel via the system than by other means. In the wilderness, traversing rivers and gorges is commonplace. In the urban environment, such a system is typically constructed when aerial companies are unable to reach high enough on tall buildings.

Traversing evolutions can be arranged more or less levelly—i.e., between two points that are roughly equal in height or between points of different elevation. Traversing by definition implies movement across the horizontal plane; however, with some additional equipment and techniques, you can work between different elevations as well. All of these evolutions are time-consuming and technical. Their inherent danger lies in the rope or series of ropes than must be tensioned between two points. The tensioning process can overstress components to such a degree that failure results. Supervised training at ground level is critical before operating at any height.

One of the primary points of contention is how much tension should be on the rope. Remembering the effects of stress from different angles on anchor points, we strive to keep the angle between the two points below 90 degrees, if possible. Otherwise, the stress to the system components is increased dramatically. The smaller the angle, the lower the stress, but with the result that the efficiency of the system is reduced. It requires more work to traverse a rope that has a lot of slack in it. See the chart on page 150.

OPERATING CONCERNS

Of primary concern is the distance to be covered. The rescue rope many fire departments carry tends to be limited to lengths of about 165 to 200 feet. Depending on how far from the edge your anchors are, plus other system requirements, longer sections are in order. This isn't an evolution in which tying ropes together is an option. Having available lengths of 300 to 600 feet is a benefit. Static rescue rope is preferred because of its low-stretch qualities. The minimum diameter is 1/2 inch, but 5/8 inch is often used for that extra margin of safety. The first static kernmantle rope I used after the original rope standard came out was the 5/8-inch for tyrolean traverses. I still like the 5/8-inch, even though half-inch is quite popular, particularly for long spans.

Depending on equipment, personnel, and the characteristics of the operating area, you can work from one end of the rope or from both ends. Having firefighters at both stations means more flexibility in performing the various tasks required by the system. One-ended operating means sometimes having support crews and management at a distance, as well as change-of-direction pulleys at the far end of the line. Suitable anchor points and access for personnel to attach to the system are no less vital to the remote stations of one-ended traversing operations.

Sometimes it is necessary to reach a victim beneath the traversing system. In such a case, a movable control point along the horizontal main line can support a vertical rope arrangement for vertical rescue.

CONSTRUCTION VARIABLES

Setting up a traversing system can be time-consuming, since it entails tackling the very problem that inspired the operation—i.e., erecting a pathway across a void. You must work out a mechanism to get a tag line across, which you will then use to pull a tensioned main line into place. When confronted by a river and valley, swimming or floating that first line across may be the best option. In the urban environment, dropping two ropes from high positions, mating them, and pulling them to one end is common.

How many tensioned ropes do you need to construct the system? The main traversing line can be a single or double line. A double line can be tensioned in direct contact or as separate units. The double-line-in-contact method usually entails a single piece of rope stretched across a gap to a change-of-direction pulley, then stretched back to its original starting point. Because this arrangement usually requires a

An Example of a Single-Rope Traversing System

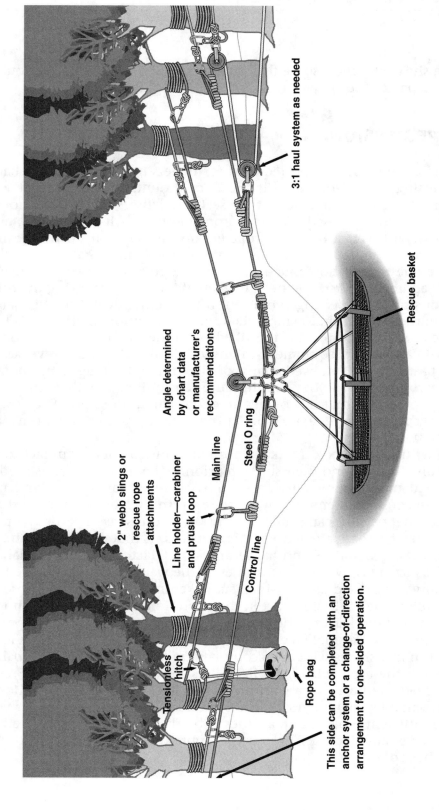

3:1 haul system as needed

Angle determined
by chart data
or manufacturer's
recommendations

Steel O ring

Main line

2" webb slings or
rescue rope
attachments

Line holder—carabiner
and prusik loop

Control line

Tensionless
hitch

Rope bag

Rescue basket

This side can be completed with an
anchor system or a change-of-direction
arrangement for one-sided operation.

wide-sheaved pulley, some teams use a single line or keep the double lines separated so they can use rescue-grade pulleys.

ROPE TENSIONING

The amount of tension required for traversing is subject to debate. Traversing came from the discipline of mountaineering, and it was then adapted and upgraded for use by the fire service. The mountaineers primarily used 7/16-inch rope because of its abundance and light weight. When half-inch rope became available, it slowly became more commonplace for these activities in the United States.

The question of how much slack to have in a tensioned rope needs to be addressed. First, you need to evaluate why you're taking the time to construct a traversing evolution. If a lowering evolution will serve the purpose, then that is most likely the best option. Setting up a traverse in ideal conditions with skilled personnel can take several hours, and it can be very equipment-intensive. While setting up traverses in the past, I have followed many of the principles taught by mountaineers—a methodology of minimum tensioning. What's good about this approach is that it provides an extra margin of safety. The downside is that it requires extra work at the distant side to recover the load being transported.

In writing this book, I talked with many instructors, manufacturers, and distributors, and the issue of tensioning became as clear as mud. I received many ambiguous answers based on urban myths, and my concern grew immensely. I was told everything from "No more than what one person can pull" to "1,000 pounds as measured on a dynometer." There are mathematical equations that can be used; however, they generally stand alone and don't factor in other variables such as ice, rain, wind loads, and even the weight of the rope itself.

Many in the field use half-inch static kernmantle for this purpose. If you follow the 15:1 safety factor, then you shouldn't load the common half-inch static kernmantle rope to more than 600 lbs. Suppose you were to tension a single section of half-inch rope to 500 lbs., then load a victim in a stokes with an accompanying firefighter. You're probably stressing the rope with 1,100 lbs. The safety factor goes from 15:1 to 8:1, even before factoring in any other local safety considerations. Such variables include how far it is across to each point; whether you have sufficient types and quantities of hardware and software; whether you have sufficient personnel; the approximate load data; and the weather conditions.

As you can imagine, the major rope manufacturers and distributors I spoke with said you shouldn't load their rope past the 15:1 safety factor. Tensioning the main line serves three main purposes. First, it minimizes the effort required to get from one point to another. Second, it maximizes rope lengths. Third, it serves to maintain clearance around any obstructions. Tensioning must be completed prior to placing any human loads on the system. Obviously, once the system is loaded with the human load, it will dip from the horizontal plane. This is where the system must clear any obstructions to be effective. And again the question arises: How tight do you pull the main line to accomplish this?

Kyle Isenhart of Rescue Systems Inc. of Lake Powell, Utah, developed a computer program that analysed many of the factors previously discussed. He came up with a table for rigging the 5/8-inch static rope (see page 266). His table is set up specifically for the Blue Water Superline rope. It follows the methodology of minimum tensioning that I was taught at the beginning and that has worked well thus far.

The chart is self-explanatory; however, I should point out a couple of things. The load to be suspended is listed in 200-, 300-, 400-, and 600-lb. increments. For one-person loads, you should look toward the 300-lb. range and to 600 lbs. for a two-person load in keeping with the rope standard. When you look at the 600-lb. tensioning requirements, the rigging tension goes up for the longer span instead of decreasing as in the other cases. This is due to the increased weight of the rope involved.

Anchoring and tensioning the ropes requires strict attention to detail. Although there are many ways to attach ropes to an anchor, the best method is to use the tensionless hitch, if possible. The process of tensioning after you have calculated what's required for the span and load can be done in several ways. A piggybacked 3:1 mechanical advantage system with two firefighters pulling or a 4:1 mechanical advantage system with one firefighter pulling should be sufficient for the higher-end tensioning. Three firefighters pulling the main line rope should be able to exert more than 300 lbs., which would work well on the lower-end tensioning. Any tensioning performed prior to the final tie-off to the far-side anchor can be held in place with rope grabs and load-releasing hitches. Do not consider any rope grabs used for this purpose to be the primary tensioning bearers after the final tie-off is complete.

SAFETYING THE SYSTEM

Regardless of whether you use a single- or a double-rope main line, you need to be able to traverse the system. This is done by using tag

TELPHER LINE-RIGGING CONDITIONS
Telpher Line-Rigging Conditions for 5/8-inch BW-Superline
Data generated by Rescue Systems Inc., Lake Powell, Utah

The data for this table is based on the two rigging points of the telpher line being horizontal and the load in the exact center, which is the worst possible case. As the load moves from the center, or as the rigging points move from the horizontal plane, the tension in the loaded telpher line will be less than is shown in the table.

How to Use This Table
1. Determine the span to be crossed.
2. Determine the load to be placed on the telpher line.
3. The third table column will tell you the proper rigging tension to achieve the prescribed conditions. The remaining columns give the dip at the center of the telpher line both before and after the load is placed on it.

Maximum tension on the rope when telpher line is loaded: 850 lbs.*

Load to Be Suspended on Rope (in pounds)	Span to Be Crossed (in feet)	Rigging Tension (in pounds)	Dip at Center of Telpher Line Before It Is Loaded (in feet)	Dip at Center of Loaded Telpher Line (in feet)
200	100	660	0.2	6.1
200	200	655	0.7	12.4
200	300	645	1.6	19.4
200	400	640	2.3	26.0
200	500	630	4.6	33.2
200	600	620	6.7	40.7
300**	100	425	0.3	9.1
300**	200	410	1.1	18.5
300**	300	400	2.6	28.2
300**	400	390	4.8	38.3
300**	500	380	7.7	48.6
300**	600	370	11.3	59.2
400	100	100	1.2	12.3
400	200	105	4.5	24.9
400	300	115	9.2	37.8
400	400	125	15.1	51.1
400	500	140	21.7	64.6
400	600	150	29.2	78.5
600	100	11.4	11.8	19.1
600	200	22.3	24.2	38.5
600	300	32.8	37.2	58.4
600	400	43.0	50.7	78.7

* 850 lbs. is the minimum safe working load for 5/8" Superline Plus based on the NFPA 1983 life safety rope standard (safe working load shall be calculated using a 15:1 ratio).
** 300 lbs. is the NFPA standard weight for a one-person load.

NOTICE
This data is for 5/8-inch-diameter Blue Water Superline rope only. It is not applicable to any ropes of other diameter or manufacture and is shown only for representational purposes. Because of the extremely complex physics of telpher-type operations, do not attempt to adapt this data.

Loose Tensionless Hitch and Mechanical Advantage

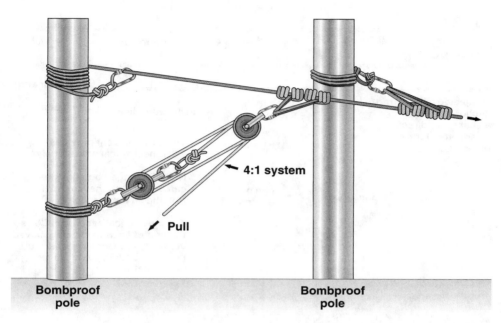

4:1 system

Pull

Bombproof
pole

Bombproof
pole

lines from either or both ends. If you run everything from one end, you'll need a change-of-direction pulley at the far end so that the tag line can be managed in both directions. Operating a tag line from both ends requires firefighters coordinating their efforts to maintain proper rope tension at all times. Tag lines also serve as a system backup and therefore must be properly anchored and secured.

WORKING IN THE VERTICAL PLANE

Moving horizontally is the main purpose of any traversing system; however, the vertical plane can be accessed with the addition of a movable control point. Essentially, the descending firefighter is attached to a mechanical advantage system that's controlled from one end of the line and which is part of the horizontal control mechanism. The firefighter is positioned over the intended target and held in place horizontally by tying off the tag lines. Members then lower him via the vertical mechanism. Exercise extreme caution in such maneuvers so as not to overtax the main system.

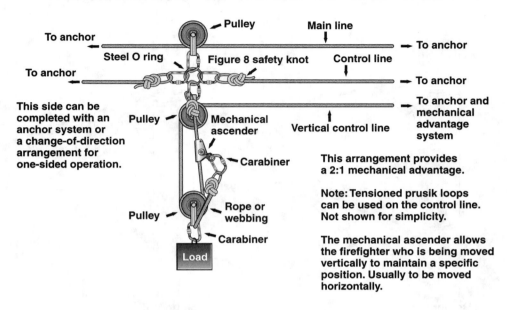

Movable Control Point in the Vertical Plane (Closeup)

To anchor — Pulley   Main line

To anchor   Steel O ring   Figure 8 safety knot   Control line   To anchor

To anchor   To anchor

This side can be completed with an anchor system or a change-of-direction arrangement for one-sided operation.

Pulley   Mechanical ascender   Vertical control line   To anchor and mechanical advantage system

Carabiner   This arrangement provides a 2:1 mechanical advantage.

Note: Tensioned prusik loops can be used on the control line. Not shown for simplicity.

Pulley   Rope or webbing   Carabiner

Load

The mechanical ascender allows the firefighter who is being moved vertically to maintain a specific position. Usually to be moved horizontally.

HIGH-TO-LOW CONFIGURATIONS

Traversing evolutions from higher to lower locations have limited use in the mainstream fire service, except for the possibility of evacuating high-rise structures. If you use this method, it must be because you have eliminated simpler lowering techniques. Given our modern, litigation-conscious culture, be sure your decision is justified!

This arrangement follows many of the operating principles of the horizontal evolution. The individual being lowered will slide down the rope using the same pulley and tag line. The tag line is a lower belay that is run through a friction device, which in turn is affixed to an anchor. You can also use such a system to transport personnel from lower to higher positions. Rather than lowering a firefighter by means of a friction device, you raise him using a mechanical advantage system located at the high end of the line. The bottomside tag line serves as a safety backup and must be properly anchored.

How much rope do you need? To figure out the minimum lengths required for a given evolution, use the Truckie Triangle. This formula is used by truck companies to evaluate whether their aerials are going to reach a building; others will recognize it as the Pythagorean theorem:

$$A^2 + B^2 = C^2$$

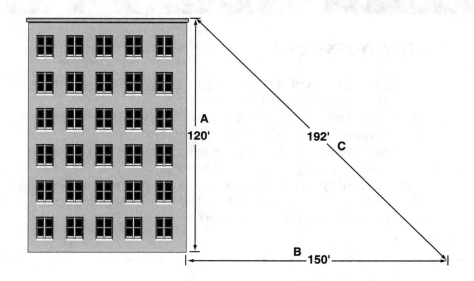

—in which A equals the vertical distance from the top of the build-ing to the ground, B equals the distance from the anchor point to the foot of the building, and C equals the required rope length. Suppose:

> *A = 120 feet, 120 X 120 = 14,400;*
> *B = 150 feet, 150 X 150 = 22,500;*
> *14,400 + 22,500 = 36,900;*
> *C = the square root of 36,900, or 192 feet.*

The answer, 192 feet, is theoretical; in actual practice, somewhat more is needed for tying off and to cover the distance between the edges and the anchor points.

Note that if the sum of $A^2 + B^2$ equals 10,000 or less, a 100-foot aeri-al ladder will have a chance of reaching the desired location on the building.

STUDY QUESTIONS

1. Traversing evolutions are also known by what three other common terms?
2. Does slack in the main line of a traversing evolution increase or decrease the stress on the system?
3. For one-ended traversing operations, a single rope requires a _____ on the far-side anchor point.
4. Tensioning the main line serves what three purposes?
5. Must tensioning be completed prior to putting any human loads on the system?

Glossary

Abrasion: The friction damage caused by two objects coming in contact, as in a rope being pulled over an unprotected edge.

Anchoring: The overall system and principle of securing yourself and equipment to a stationary object.

Anchor point: A substantial object that you select to be a load-bearing component of an anchor system.

Anchor system: A complete load-bearing assembly, comprised of one or more anchor points, rope, webbing, and hardware.

Artificial anchor point: A man-made anchor point located so as to be convenient to the system.

Ascender: One of several devices that allow an individual to climb a vertical rope or that serve as a rope grab.

Assignment: A mission or set of tasks for an operation.

Back up: The act of safetying a system.

Backup system: An independent, redundant system constructed and in place to meet the demands of the operation in the event of failure of the primary system.

Becket: A carabiner attachment point on the bottom of a double pulley.

Belay: A rope that secures a climber to an anchor point as a safety; also, the act of securing a climber in this way.

Belayer: The team member responsible for managing the belay line and friction device to protect the suspended rescuer.

Belay plate: A friction device used as a brake to dampen the force of a falling member.

Bight: A rope that is bent back on itself and parallel.

Bottom belay: A safety line used in single-rope techniques with which a team member at a lower position can arrest the fall of a climber.

Brake bar rack: A U-shaped metal device employing bars through which the rescue rope is reeved and which is used to provide friction during rope evolutions.

Brake hand: The hand used during rope descents to maintain control by gripping the rope.

Cam: A rope grab.

Carabiner: Metal connecting links used extensively in rope systems.

Chest harness: A load-supporting life safety device worn around the upper torso; usually worn where inverting can occur or for retrieval from confined spaces.

Company: The basic unit of fire service members under the control of an officer.

Compound mechanical advantage: The multiplying of mechanical advantage by stacking simple systems together.

Confined space: Any space that a person can get into but which is not designed for continuous human habitation. See OSHA 29 CFR 1910.146.

Cordage: Rope, webbing, and accessory cord.

Descender: A device that provides friction to control the rate of descent on a rope.

Directional: The quality of a pulley to produce a more favorable operating position by altering a system.

Double-rope technique (DRT): A primary rope system backed up by an independent and redundant secondary system.

Dress up: To neaten and check the accuracy of a tied knot.

Edge: Any surfaces that form corners for rope and from which rope and equipment must be protected.

Edge protection: A wide variety of soft and hard devices to protect moving ropes from abrasive surfaces.

Elongation: The stretching of a rope, usually expressed as a percentage.

Emergency egress operation: Any expedient exit from a dangerous situation, often using nontraditional methods.

Engine: The basic apparatus for fire suppression, comprised of a pump, a water tank, and a hose supply.

Environment: In the context of rope rescue, the general operating area and its features.

Evolution: A specific task or group of techniques within an operation.

Fall factor: The resultant value assigned after computing the length of a fall vs. the amount of rope that has been paid out in a system.

Figure 8: A common descending device shaped like an eight.

Fixed brake: A friction device placed in a nonmoving section of a rope system.

Full-body harness: A personal protective device worn around the upper and lower torso.

Guide hand: The hand that grasps the vertical rescue rope for stability while descending.

Hardware: An inclusive term for the metal connecting, grabbing, and friction devices and other components of a rope system.

Haul cam: The rope grab that provides gripping action during a hauling evolution.

High angle: Denotes operations in which members are dependent on life safety rope and not a fixed surface for support.

Incident command system (ICS): Operational heirarchy and guidelines for officers to control the various aspects of an incident.

Independent: In rope rescue, a redundant system that is separate from the main load-bearing system but which is equally as strong.

Kernmantle: A rope design featuring an inner, stronger core covered by an outer sheath.

Ladder or truck: An apparatus equipped with an aerial device.

Laid rope: Rope constructed by twisting fibers into three large strands that are then combined into one rope.

Life-safety system: Any mechanism comprised of several subsystems that directly supports a life during a rope rescue.

Load: The common term for the total weight of personnel and equipment being supported by a rope system.

Low angle: Denotes operations in which firefighters are dependent on the ground for their primary support and rope systems for their secondary means of support.

Lowering: The technique of moving a load from a high to a low position; if the load is an individual, that person has no direct control over the descent.

Main line: The primary load-bearing rope system in a double-rope system.

Needle loom: The latest construction technique used to manufacture rescue-grade webbing.

NFPA: The National Fire Protection Association. This organization, based in Quincy, Massachusetts, develops operational and safety standards for industry and the fire service.

Operation: The overall coordinated effort by sanctioned forces to mitigate an emergency.

OSHA: The Occupational Safety and Health Administration, a federal agency charged with overseeing employee safety in the workplace.

Rappelling: The common term for personal descending techniques.

Ratchet cam: The rope grab that holds the load while the mechanical advantage system is being reset.

Rescue company: A group of firefighters that are also trained in specialized rescue techniques, such as rope operations.

Rescue load: A load that doesn't impose more than 600 lbs. on a rope system.

Rope grab: A mechanical device or prusik knot tied to the rescue rope that functions as a gripping mechanism.

Rope rescue: Any operation where a life-dependent rope evolution is required.

Running end: The end of the rope that is to be used for work such as hoisting or pulling.

SAR: Search and rescue; also supplied air respirator.

SCBA: Self-contained breathing apparatus.

Shuttle loom: The preferred construction method for tubular webbing used by those in rope rescue. Also known as spiral weave, it is being replaced by needle loom.

Simple mechanical advantage: Gained by noncomplex systems, usually one rope and several pulleys that work together as a unit.

Single-rope technique (SRT): The use of a primary rope system without a redundant secondary system—i.e., one rope.

Software: The soft components of a rope system, such as rope, webbing, and accessory cord.

Standing part: The section of rope between the working and running end.

Traveling brake: A friction device in a moving position, which usually means attached to the descending firefighter.

Working end: The part of the rope that forms the knot.

Answers to Study Questions

Chapter One
1. Topography, construction/occupancy, demographics, and in-house capabilities.
2. Fire resistive, noncombustible, ordinary, heavy timber, wood frame.
3. The purpose or use of a given building.
4. 29 CFR 1910.146, Permit Required Confined Space Regulation.
5. Demographics.

Chapter Two
1. Operations.
2. Strategy.
3. Locate, Access, Stabilize, Transport.
4. SARA, the Superfund Amendments and Reauthorization Act of 1986.
5. Command, Operations, Planning, Logistics, and Finance.

Chapter Three
1. Psychomotor.
2. Class III.
3. Hardware and fabric.
4. Abrasive and sharp edges.
5. Confined spaces.

Chapter Four
1. NFPA 1983, Fire Service Life Safety Rope and System Components.
2. Block creel construction.
3. Nylon and polyester.

4. (a) It stretches under load, (b) it untwists when loaded and causes a spinning or kinking reaction, (c) the rope's integrity can be compromised when its load-carrying fibers are exposed to the environment, abrasive surfaces, and heat-generating mechanical devices, (d) nylon hard-laid ropes are difficult to manage for transport and deployment because their stiffness limits your options, and (e) more friction is created by laid ropes due to their bumpy exterior.
5. Static kernmantle.

Chapter Five
1. (a) General use, (b) 9,000 lbs.
2. Major or longitudinal axis.
3. Earred figure 8.
4. Brake bar rack.
5. (a) To gain mechanical advantage, (b) to change the direction of the rope, and (c) to reduce or eliminate the rope's contact with an abrasive surface.

Chapter Six
1. Bend.
2. Bowlines, figure 8s, and hitches.
3. Basic figure 8 (figure 8 stopper knot).
4. As a rope grab.
5. 4:1.

Chapter Seven
1. Tensionless, equalizing, nonequalizing, and multiple.
2. Man-made improvised.
3. 70 to 75 degrees.
4. 3-2-1.
5. Frames, axles, bumpers, and rated tow hooks.

Chapter Eight
1. 600 lbs. (272 kg.)
2. Tensionless anchor system.
3. 90 degrees.
4. Anchor plate.
5. Multiple anchor point system.

Chapter Nine
1. Length of rope that is out in the system.
2. 12 kN.
3. Dynamic kernmantle.
4. The rope.
5. Running protection.

Chapter Ten
1. The higher ground or above.
2. Short board.
3. (a) It provides for easier handling of the victim, (b) it helps to prevent further injury by stabilizing the entire body, (c) it is typically used with a short board, and (d) the two-board method can facilitate moving a victim through a small opening.
4. Stokes basket.
5. Secondary backup (belay).

Chapter Eleven
1. Cavers.
2. Rope, Evolution, Systems, Consequences, User, Equipment, Support.
3. Braking.
4. On top.
5. Away.

Chapter Twelve
1. Vertical block system.
2. Piggyback.
3. Even.
4. The weight of the load and the effort being expended.
5. 5:1.

Chapter Thirteen
1. (a) Descending, (b) lowering.
2. Parallel.
3. High-angle evolution.
4. Hard and soft.
5. Bottom.

Chapter Fourteen
1. SRT.
2. Class II.
3. Three.
4. Top.
5. Pushing with your legs.

Chapter Fifteen
1. Tyroleans, telphers, and high lines.
2. Decrease.
3. Change-of-direction pulley.
4. (a) It minimizes the effort required to move from one point to another, (b) it maximizes the length of the rope, and (c) it maintains clearance around obstructions.
5. Yes.

Endnotes

Chapter 1, page 14
[1]Pedrotti, Dean R., High-Angle Rescue, Phoenix FD's Technical Rescue Team, New York: *Firehouse* Magazine, November 1990.

Chapter 2, page 27
[2]Setnicka, Tim J., ed. Wilderness Search and Rescue, Boston: Applachian Mountain Club, 1980, p. iii.

Bibliography

Beers, D.E. and Ramirez, J.E., 1990, "Vectran Fiber for Ropes and Cables," Charlotte, NC: Hoechst Celanese Corporation.

Brown, Patrick, 1991, "FDNY: Two Rescues in the Sky," *Fire Engineering,* July 1991, Saddle Brook, NJ, PennWell Publishing.

CMC Rescue, Inc., 1997, Product Catalog 118, Santa Barbara, CA: California Mountain Company, Inc.

Department of the Army, 1976, "Military Mountaineering," Washington, D.C.: Headquarters Department of the Army.

DBI/SALA, 1993, Product Catalog, Red Wing, MN: DB Industries, Inc.

DuPont, 1993, "DuPont Fibers, Bound for Performance," Wilmington, DE: DuPont, Inc.

DuPont, 1992, "Kevlar Aramid," Information Bulletin, Wilmington, DE: DuPont, Inc.

DuPont, 1993, "Properties of DuPont Industrial Filament Yarns," Wilmington, DE: DuPont, Inc.

DuPont, 1991, "Rope and Cordage Safety," Wilmington, DE: DuPont, Inc.

FEMA, 1989, "Incident Command System," Emmitsburg, MD: Federal Emergency Management Agency: National Emergency Training Center: National Fire Academy.

FEMA, 1993, "Rescue Systems 1," Emmitsburg, MD: Federal Emergency Management Agency: National Emergency Training Center: National Fire Academy.

Frank, James A. and Smith, Jerold B., 1992, "Rope Rescue Manual," 2nd edition, Santa Barbara, CA: CMC Rescue, Inc.

Hawill's Limited, 1996/1997, Product Catalog, Westborough, MA: Hawill's Limited.

IFSTA, 1992, "Essentials of Firefighting," 3rd edition, Stillwater, OK: Fire Protection Publications: Oklahoma State University.

IFSTA, 1981, "Fire Service Rope Practices," 5th edition, Stillwater, OK: Fire Protection Publications: Oklahoma State University.

MFT&E, 1995, "Firefighter 1 1995 Update," South Portland, ME: Maine Fire Training and Education.

NCFSA, n.d., "Confined Space Rescue," Bethpage, NY: Nassau County Vocational Education and Extension Board.

NFPA, 1995, "1983 Fire Service Life Safety Rope and System Components," Quincy, MA: National Fire Protection Association.

NHFST, 1981, "Training Bulletins Level II Fire Fighter," Concord, NH: New Hampshire Fire Standards and Training.

OSHA, 1995, "Fall Protection in Construction," Washington, D.C.: U.S. Department of Labor: Occupational Safety and Health Administration.

OSHA, 1993, "29 CFR 1919.146 Permit-Required Confined Spaces for General Industry Final Rule," Washington, D.C.: U.S. Department of Labor: Occupational Safety and Health Administration.

Padgett, Allen and Smith, Bruce, 1987, "On Rope," Huntsville, AL: National Speleological Society.

Petzl International, 1996/97, Product Catalog, Crolles, France: Zone Distribution, Inc.

PMI-Petzl, 1997, Work and Rescue Product Catalog, Lafayette, GA: PMI-Petzl Distribution, Inc.

Rescue Systems, Inc., 1997, Product Catalog, Lake Powell, UT: Rescue Systems, Inc.

Rescue Technology, 1997, Product Catalog, Carrollton, GA: Rescue Technology.

Setnicka, Tim. J., 1980, "Wilderness Search and Rescue," Boston, MA: Appalachian Mountain Club.

Smith Safety Products, Inc., 1997, Product Catalog, Petaluma, CA: Smith Safety Products, Inc.

Sterling Rope Company, Inc., 1996, Product Catalog, Beverly, MA: Sterling Rope Company, Inc.

The Mountaineers, 1985, "Medicine for Mountaineers," Seattle, WA: The Mountaineers.

Vines, Tom and Hudson, Steve, 1989, "High-Angle Rescue Techniques," Fairfax, VA: National Association for Search and Rescue.

Wheelock, Walt, 1967, "Ropes, Knots, and Slings for Climbers," Glendale, CA: La Siesta Press.

Index